Economics of Brackishwater
Shrimp Culture

Palaparthi Andrew

2018

Daya Publishing House®

A Division of

Astral International Pvt. Ltd.

New Delhi – 110 002

Reprint 2018

ISBN: 978-81-7035-216-7 (HB)

Published by : **Daya Publishing House®**
A Division of
Astral International Pvt. Ltd.
– ISO 9001:2015 Certified Company –
4736/23, Ansari Road, Darya Ganj
New Delhi-110 002
Ph. 011-43549197, 23278134
E-mail: info@astralint.com
Website: www.astralint.com

Digitally Printed at : Replika Press Pvt. Ltd.

Dedicated

To

My Beloved Parents

Foreword

In an agro-based economy like that in India, agriculture is the chief source of employment and income. But it is observed that the income generated through this sector is so meagre that it fails to satisfy even the basic needs of the rural population. In the recent times ancillary and allied, activities like dairy, poultry sericulture, horticulture, Aquaculture' etc., have attracted considerable attention of the rural people. Aquaculture in particular plays quite a significant role in the economic life of the rural poor, especially in the coastal areas. Although fishing is an age-old caste based activity, Aquaculture of shrimp has emerged as a highly commercialised and scientific activity. It has become a prime investment area as it steps up foreign exchange earnings besides providing food and employment. The rate of returns and productivity in aquaculture is proved to be higher than many major agricultural crops in India.

'Shrimp' which is an aquatic organism inhabiting the sea, estuaries and backwaters is the backbone of Indian marine products export. The value of shrimp exported form India annually is estimated at nearly Rs. 3,000 crores. India is now recognised as one of the world leaders in shrimp farming. The Government of India identified shrimp farming as 'thrust area' for export promotion because of its persistent demand in the global market and vast potential for development. Due to the unstinted organisational support given by the government, the pace of development of brackishwater shrimp culture has been impeaceable and fairly rapid over the past few years. Some of India's large private companies have already invested sizeable amounts of money in setting up shrimp farming projects with high future expectations. While the progress in shrimp farming is quite essential, the policy making would be guided by indepth awareness of the implications of commercial shrimp culture. It ensures the growth of shrimp culture and averts any adverse consequences on the environment.

Dr. P. Andrew's study on 'Economics of Brackishwater Shrimp Culture' is a significant contribution towards the increasing awareness of the economic, social and environmental implications of brackishwater shrimp culture as an allied agriculture activity.

Dr. Andrew's case study has three distinct aspects. Firstly, his selection and survey of shrimp culture farmers of Krishna District of Andhra Pradesh on the basis of its significant place in shrimp culture, the availability of farmers belonging to different castes and categories and their experience in shrimp farming. Secondly, an excellent account of gensis, evolution of aquaculture in general and brackishwater shrimp culture in particular. In the third place the socio-economic characteristics of shrimp culture farmers are well presented. His case study though limited to three mandals of the district, is based on detailed enquiries into several aspects of shrimp culture such as costs of production, returns on investment, employment generation, marketing issues etc. As is to be expected, the aquaculture of shrimp is a lucrative and income generating activity. The study notes caste is not a bar for people in the coastal villages to take up shrimp farming. Another interesting observation of this study is that the corporate firms are making inroads into commercial shrimp culture activity.

Dr. Andrew's assiduous interdisciplinary case study of brackishwater shrimp culture is preceded by an interesting analysis of the significance of the shrimp culture in the national economy, the gensis and evolution of aquaculture and also by a review of number of published and unpublished studies on brackishwater shrimp culture. It is undoubtedly a valuable study that would add to our knowledge on the economics of brackishwater shrimp culture.

C.S.N. Raju, M.A., Ph.D.,
Professor
Department of Economics,
Nagarjuna University,
Nagarjuna Nagar (A.P)

Preface

The purpose and the scope of this book may easily be gathered from the introductory chapter. Aquaculture of shrimp, still being a young science with enormous potential for innovation, a few words of preface are needed.

India has been recognised as one of the world leaders in cultured shrimp production with its scintillating performance in the last couple of years. Although the allied agriculture activity like 'Aquaculture' had ancient traditions, it has emerged as a promising solution to the problem of food shortage and malnutrition only in recent times. Moreover, by bringing into use the vast derelict water spread areas, it provides employment to a number of people in rural areas and earns substantial foreign exchange to the exchequer.

In India, scientific aquaculture has become synonymous with brackishwater shrimp culture which gained its pace of development in the nineties and has turned out be quite lucrative. Hence it has drawn the attention of the industrial world and has become a major topic of interest and discussion. The short gestation period, high returns, ready market and vast scope for future development are some of the features of this enterprise.

In the seafood exports from India, shrimp has been a dominant item. With depleting catching quantum of shrimp from capture sources, the accent has rightly been on brackishwater shrimp aquaculture. The best species for culture in India are Tiger shrimp or Prawn (penaeus monodon) and White shrimp (p. indicus). These two constitute 75 per cent and 20 per cent respectively while the other species constitute only 5 per cent. There are four types of shrimp farming viz., Traditional, Extensive, Semi-intensive and Intensive type. While the small and marginal farmers are using the extensive type the large private farms are adopting the semi-intensive type.

The Ministry of Commerce of Government of India has Identified shrimp culture as thrust area for export promotion and brought it under 'Extreme Focus Category'. The new industrial policy and export-import policy announced by the government in the early nineties and the duty exemption given for the equipment and components of shrimp culture added impetus to boost the aquaculture establishments.

Brackishwater shrimp culture has emerged as a multi-disciplinary science starting with Biology and attracting many sciences into its fold during the last one and a half decades, the latest being Economics.

While the increase in the production of shrimp is clearly visible from the data available, the economic aspects of brackishwater shrimp culture like the employment and income generation, marketing issues, the socio-economic profile of shrimp farmers etc. have not been adequately studied. While the intensive research is needed from time to time to focus on the above issues, nothing worthwhile is available in the existing literature. The present study therefore has made a modest attempt to answer some of these questions. The study also made a sincere attempt to bring out the Genesis and Evolution of Aquaculture and its relative advantages over capture fisheries to provide a clear understanding of this miracle method.

The book is concerned only with the brackishwater shrimp culture economics in Krishna District where the 'Aquaplosion' of shrimp is said to have taken place.

P. Andrew

Contents

Abbreviations Used

A.P.A.U.	:	Andhra Pradesh Agricultural University
A.S.E.A.N.	:	Association of South East Asian Nations
A.I.C.R.P.	:	All India Co-ordinated Research Project
A.P.P.C.B.	:	Andhra Pradesh Pollution Control Board
B.F.F.D.A.	:	Brackishwater Fish Farmers Development Agency
B.O.B.P.	:	Bay of Bengal Programme
C.I.B.A.	:	Central Institute of Brackishwater Aquaculture
C.I.F.A.	:	Central Institute of Freshwater Aquaculture
C.I.F.E.	:	Central Institute of Fisheries Education
C.M.F.R.I.	:	Central Marine Fisheries Research Institute
D.R.D.A.	:	District Rural Development Agency
E.E.Z.	:	Exclusive Economic Zone
E.F.C.	:	Extreme Focus Category
E.R.R.P.	:	Economic Rehabilitation of Rural Poor
E.O.U.	:	Export Oriented Units
I.C.A.R.	:	Indian Council of Agricultural Research
I.C.IC.I.	:	Industrial Credit and Investment Corporation of India.
I.Q.F.	:	Individual Quick Frozen

M.P.E.D.A.	:	Marine Products Export Development Authority
N.A.B.A.R.D.	:	National Bank for Agriculture and Rular Development
N.E.E.R.I.	:	National Environmental Engineering Research Institute
O.S.S.P.A.R.C.	:	Orissa Shrimp Seed Production and Research Centre
P.H.	:	Potential Hydrogenii
P.P.T.	:	Part Per Thousand
R.D.	:	Research and Development
S.E.A.F.D.C.	:	Sough East Asian Fisheries Development Centre
S.C.I.C.I.	:	Shipping Credit and Investment Company of India
S.P.C.B.	:	State Pollution Control Board
T.A.S.S.P.A.R.C.	:	The Andhra Shrimp Seed Production and Research Centre
U,N.D.P.	:	United Nations Development Project
U.N.C.T.A.D.	:	United Nations Council for Trade and Development
V.A.P.	:	Value Added Product
W.T.O.	:	World Trade Organisation

Technical Glossary

Aquaculture	:	Growing aquatic organisms like fish and shellfish under certain controlled conditions.
Brackishwater	:	Any mixture of seawater and freshwater with a salinity of less than 35 parts per thousand (PPT). It is intermediate in taste between the salt and freshwater. It is also perfectly buffered against abrupt changes in PH.
Coastal Zone	:	Strip of land of variable width along the shoreline under the direct environmental and economic influence of the adjoining body of water.
Commission Agent	:	One who purchases bulk of shrimp from the farmers on behalf of the processor or exporter for commission.
Continental Shelf	:	The seabed and subsoil of the submarine areas adjacent to the coast, to a depth of 200 metres or beyond that limit, to where the depth of the super-jacent waters admits to the exploitation of the national resources of such water.
Crop Holiday	:	Refers to a period of temporary gap of three to four months deliberately given in aquaculture operations by the farmers to counter a viral disease.
Creek	:	Small stream—a narrow inlet on seacoast or lakeshore.

Depreciation	:	Depreciation on fixed capital is calculated as follows : Depreciation = $\frac{\text{Newly purchasedvalue-Scrap Value}}{\text{Economic Life of the Asset}}$
Ecology	:	Field of science that deals with the relationship between living organisms and their environment.
Environment	:	Sum of physical, biotic social factors and conditions directly or indirectly affecting the development, life and activities of organisms and populations in the short and long terms.
Estuary	:	A place where a river empties into open sea (river mouth), forming a bay subject to tidal effects, where freshwater and seawater mix.
Fish	:	Includes fin-fish, mollusks and crustaceans.
Fingerling	:	Young or small fish or shrimp.
Fry	:	Small fish or recently hatched fish or shrimp.
Hatchery	:	A place where eggs especially those of fish or domestic fowl are hatched i.e., to produce from eggs.
Interest	:	Interest on capital in the study is calculated at the rate of 18%.
Lagoon	:	Salt water lake cut off from the sea.
Labour	:	Human labour both family and hired labour engaged in shrimp culture.
Pollution	:	Emission of pollutants-physical, chemical or biological substances resulting from human activities that reduce the quality of the environment.
Pokkali	:	A variety of paddy or rice field of Kerala.
Potential Hydrogenii	:	It can be identified as the negative logritham of the concentration of hydrogenions in moles per litre.

Salinity	:	Refers to the amount of salt in water expressed in parts per thousand(PPT).
Satellite Farming	:	Principal farms or corporate farms set up their own farms, hatchery, feed mill, R & D Unit and processing plant not only for their own needs but also to cater to the needs of many small farmers around them is called satellite farming.
Shrimp Farmer	:	The person who is engaged in shrimp farming.
Mangrove	:	A tropical maritime tree that sends out many prop roots and forms dense thickets important in coastal land buildings.
Non-recurring cost or Capital Investment	:	It covers construction of main and subsidary bunds/dikes, sluice gates, excavation of ponds, construction of watchman shed-cum-store house, cost of equipment such as, pump sets, areators, nets, screens, shutters, etc.
Recurring cost or Operational Cost	:	It includes cost on pond preparation, seed, feed, fertilizers, chemicals, fuel, wages of workers, repairs of bunds and equipment, etc.
Vertically Integrated Aquaculture	:	If an aquaculture farm is having its own hatchery, feed mill, R & D Unit and processing plant, etc., it is called vertically integrated aquaculture.

List of Tables

Chapter 1

Introduction

Indian economy is predominantly an agro-rural economy. A cursory glance at the rural economy of India generally presents the portrait of a number of people living under conditions of poverty and penury, dwelling in sequestered settlements and wearing a morose and lackadaisical look. This segment of the Indian economy is characterised by inadequate momentum and motivation necessary for economic upliftment and emancipation. Therefore, the rural sector in totality appears to be a spectrum of incohesive and disjointed block of immense but inert resources and potentialities.

Economic justice to the people is inseparable from political freedom. Freedom is not meaningful and complete if it does not provide a better deal and opportunity for the poor and neglected strata of the society in general, and rural component of it in particular. The major malady of rural scene in India, as identified by Mahatma Gandhi, has been the divorce between intellect and labour which resulted in condemning millions of persons living in despair and drudgery, especially in the villages.[1] It is true that, the research conducted and the contributions made by Indian scientists and technocrats in the field of science and technology is mostly confined to urban-industrial sector. The existing rural socio-economic scenario therefore has not significantly changed from what it has been in the past.

Although India has registered impressive achievements in the economic sphere after independence, the pattern of imbalance almost remains the same with the affluence of the few and the

1. Swaminathan, M.S., "Initiating a self replicating process of rural regeneration", in S.R. Subrahamanian (Ed.), *Readings in Integrated Rural Development*, Oxford and IBH Publishing, New Delhi, 1987, p. 3.

misery of many. The urban-rural dichotomy in the matter of development continues to exist. The benefits of planned development achieved so far could not penetrate and percolate into large sections of deserving rural masses. This unevenable record of phenomenal progress in certain segments and significant failures in some other segments are alarming enough to threaten socio-economic foundations of the country, especially in the undeveloped and underdeveloped rural areas.

The development of India is synonymous with the development of villages because India is a country wherein nearly seventy per cent of the population dwell at the country side.[2] Despite the constant endeavour and consistent policy of the government to improve the living conditions of the rural people right from the dawn of economic planning, there is not much change in the state of rural economy in India. Therefore, Indian efforts to establish an egalitarian society have to be viewed and reviewed from this perspective.

In an agro-based economy like India, agriculture sector is the chief source of employment and income. But the income generated through this sector is so meagre that it fails to satisfy the minimum basic needs of landless and marginal farmers who constitute a sizeable work force among the rural population.[3] The rural labour are generally paid discriminative and unremunerative wages. Under-employment coupled with varying spells of unemployment and lack of opportunities for alternative vocation and income sources are quite common with these people. It is observed that in the Indian context agriculture along cannot weed out unemployment and generate adequate income to relieve the rural poor from the clutches of object poverty and squalor. The Indian rural scene thus presents a complex milieu making the task or rural development hard and formidable. The rural development has become a key factor in the alround economic progress of the country. It is therefore imperative that providing all possible opportunities to the rural poor for generating additional employment and income should be an integral part of comprehensive long term economic strategy of the government.

---

2. Sharma, D,P., and Desai, V.V., *Rural Economy* of India, Vikas Publishing House, Delhi, 1980, p. 5.
3. Radhashyam and Tripathy, N.K., "Aquaculture as Nucleus for Integrated Rural Development—An Experience", *Fishing Chimes*, Dec., 1992, pp. 37-41.

The main purpose of this introductory chapter is to emphasise the role of aquaculture, especially brackishwater shrimp culture in the rural development of India as an allied activity for agriculture. It throws light on the significance of the study area in brackishwater shrimp culture, the objectives and the methodology of the present study.

Importance of Aquaculture in India

In addition to traditional agriculture, ancillary and allied activities like dairy, poultry, sericulture, horticulture etc. have attracted considerable attention of the rural people. However, a very profitable activity like aquaculture has been ignored till very recently. Aquaculture (farming and husbandry of aquatic organisms like fin-shellfish etc. under controlled conditions) plays quite a significant role in the economic life of the rural poor. Aquaculture is said to be a prime investment area as it provides food, steps up foreign exchange earnings and creates wealth and prosperity much more than any other sector.[4] The returns from aquaculture is 2 to 15 times higher than traditional agriculture. The productivity is also higher in aquaculture *i.e.* about 25 tonnes per hectare per year, compared to 2 to 5 tonnes per hectare per year in major agricultural crops like cereals.[5]

However, several observations made on the fish through capture of natural wild stock indicate that the fish landing is either stagnant or depleted in traditional fishing areas. Compared to the sluggish growth of fisheries supply, the consumption of fishery products has been increasing in tandem with exponential growth of population, creating a huge gap between supply and demand for fisheries.[6] Aquaculture development is expected to bridge the gap between the demand for and supply of fish because aquaculture yields more fish per unit area than that of natural sources.[7]

4. Silas, E.G., "Towards sustainable Aquaculture", *Fishing Chimes*, May, 1993, p. 32.
5. Menon, K.K.G., "Basic approaches for higher aquaculture productivity", in V.R.P. Sinha and H.C. Srivastava, (Eds.). *Aquaculture Productivity*, Oxford and IBH, New Delhi, 1991, P.11.
6. Marine Products Exports Development Authority, *Hand Book on Shrimp Farming*, Kochi, 1992, p.6.
7. Srinivasa Sastry, C., Keynote Address to the National Workshop on Brackishwater Prawn Farming for Higher Production, Bangalore, Oct., 23-25, 1989, p. 4.

Although India occupies 7th position among fish-producing countries, the actual contribution of fisheries to the gross domestic product (GDP) is only 2.4 per cent. India's share in 50 billion seafood trade of the world is only 2 per cent. In this context aquaculture in general and brackishwater shrimp culture in particular is expected to be instrumental in bringing about the socio-economic uplift and rehabilitation of economically weaker sections in the villages, by providing them additional income and employment opportunities.[8]

Brackishwater Aquaculture in India

Brackishwater aquaculture is considered to be the most promising part of aquaculture in India in recent times. As such, the aquaculture scene of India is being shifted to brackishwater aquaculture where the potential for development is very high. Commercial brackishwater aquaculture in India is synonymous with 'shrimp' [9] (prawn) culture. Shrimp farming is considered to be the most lucrative enterprise due to high market price and persistent global demand. It is interesting to note that in spite of India being a minor player in the total seafood trade, the country enjoys a fairly prominent position in the world shrimp market. The country accounts for 11 per cent share of total world production of shrimp and commands 7 per cent share in quality and value. Shrimp farming development also holds immense employment potential for the local people. It provides direct employment to at least two persons per one hectare culture area. It also provides indirect employment to the extent of 1100 man days per hectare during construction, input supplies, handling of material, harvesting etc. [10]

Although the cultured shrimp is a nutritious food item, the domestic market for shrimp is quite negligible. Only peeled and second rate shrimp is available in the domestic market. This is due

8. Sandip, K. Mathur and Gopal Krishnan, G., "Economics of Commercial Brackishwater Farming : Case Studies in Andhra Pradesh", *Aquaculture Productivity* ... Op. Cit., pp. 269-280.
9. Though the words 'Shrimp' and 'Prawn' are used interchangeably in Indian context, in the present study only the word shrimp is used. 'Shrimp' is used in this study both as singular and plural. (See Appendix-I).
10. MPEDA, *Hand Book on Shrimp Farming, Op. Cit.*, p. 21.

to the fact that shrimp is exported to the international market where it fetches very high price. Therefore, brackishwater cultured shrimp is said to be 100 per cent export oriented item. In 1992-93 the total value of marine exports was estimated at Rs. 1,767.43 crores. Of this shrimp alone accounted for Rs.1,190.26 crores. In 1993-94 the value of marine exports was about Rs. 2,503.62 crores, out of which shrimp accounted for about 80 per cent.[11] Brackishwater culture of shrimp thus helps the country to earn foreign exchange which is very precious for a developing country like India.

In the brackishwater shrimp farming sector there had been a phenomenal growth in the country. Area under brackishwater farming went upto a level of 82,500 hectares against the estimated potential area of 11,90,000 hectares with a shrimp production of 62,000 tonnes per annum in 1993-94. The average production of shrimp per hectare improved to 750 Kgs. from 560 Kgs. per hectare in the previous year. In 1994 there had been a remarkable increase in shrimp hatcheries set up in the country. Around 100 units had been set up chiefly in Andhra Pradesh, Tamil Nadu, Pondichery, Maharashtra, Goa, Karnataka and Kerala recently. These units are capable of producing 4,000 million seed annually when completed.[12]

Brackishwater Aquaculture Development Under Five Year Plans

The Government of India recognised brackishwater aquaculture as one of the potential areas of fisheries development during the 5th Plan period (1974-79), and established Pilot Project Centres and Demonstration-cum-Extension Centres for its development. During the 6th Plan 'Area Development Scheme' was introduced to develop this part of aquaculture. In the 7th Plan, the government took up 'Integrated Brackishwater Fish Farm Development'. As part of this programme, the project on 'Development of Coastal Aquaculture' assisted by United Nations Development Project (UNDP) and executed by Food and

---

11. Verghese, P.U., "Aquaculture Production to augment export of marine products from India" in *Indaqua-Aquaculture Exposition, Souvenir,* MPEDA, Madras, Jan., 27-29, 1995.

12. Dixitulu, J.V.H., "Indian Fisheries Sector 1994", *Fishing Chimes,* Jan, 1995, p. 13.

Agricultural Organisation (FAO) was started with the aim of establishing 5 Pilot Fish Projects and 5 Pilot Shrimp Hatcheries.

During the 8th Plan period brackishwater shrimp culture is identified as 'Extreme Focus Area' from the stand point of export value and foreign exchange earning capacity. Marine Products Export Development Authority established by the Ministry of Commerce, Government of India, plays a catalystic role in developing brackishwater shrimp farming and marketing.

Sources of Fisheries in Andhra Pradesh

The state of Andhra Pradesh is unique in having very rich fisheries resources in both marine and inland areas. The state occupies third position in fish production, while Bengal and Bihar take first and second positions respectively. [13]

The aquatic resources of the state can be broadly divided into two types, viz., (1) Marine and (2) Inland. The inland fishery resources are further divided into capture and culture fisheries. Andhra Pradesh is the second largest maritime state in India. It has a coast line of 982 Km. and a continental shelf area of 39,109 sq. kilometers.[14] Thus the state has rich marine and inland fishery resources. They comprise two types of waters, namely, fresh water and brackishwater. The former includes river system, irrigation canals, reservoirs, lakes, tanks, ponds etc. while the latter consists of estuaries, lagoons, brackishwater lakes, backwaters and vast areas of mangrove swamps containing tidal waters of fluctuating salinity. These productive water bodies constitute the state's most valuable culture fishery asset. The brackishwater area found to be suitable for shrimp culture in the state is estimated at 1,50,000 hectares.

Brackishwater Shrimp Culture in Andhra Pradesh

Brackishwater aquaculture is a new venture in Andhra Pradesh when compared to fresh water fish culture. Brackishwater aquaculture in Andhra Pradesh is confined to shrimp culture only. Although this activity was started on a modest note in 1980s, within a short span of 10 years it made rapid strides and overtook

13. Commissioner of Fisheries, *Hand Book of Fisheries Statistics of Andhra Pradesh*, Government of Andhra Pradesh (1992-93), Hyderabad, p. 23.

14. Subba Rao, N., Economics of Fisheries : A Case Study of Andhra Pradesh, Daya Publishing House, Delhi, 1986, p. 1.

freshwater aquaculture in a aquaculture race. By 1993-94 about 19,500 hectares of brackishwater area was brought under shrimp culture and about 26,000 metric tonnes of shrimp production had been achieved in the state. Andhra Pradesh stands second in the country with 23.62 per cent in area under shrimp culture and occupies first place with 42 per cent of shrimp production in the country.

The entry of private entrepreneurs into the brachishwater shrimp culture recently has changed the complexion of the aquaculture scene of the state. The rush of private entrepreneurs into this venture commenced with the entry of Rank Aqua in 1989. Subsequently several firms like, Thapar, Waterbase, Alsamarine, Magunta, Coromandal Aqua etc. have made their entry into this 100 per cent export oriented enterprise. [15] Some of these private firms have also entered into the capital market with public issues in the state. Several foreign technical firms from Taiwan, T hail and, Philippines and Japan have also established trade and technical links with these corporate firms.

Towards promoting sustainable and increased shrimp production in the state, a project with the financial and technical assistance from the World Bank was taken up with capital outlay of nearly Rs. 83.13 crores.[16] The Government of Andhra Pradesh has set apart Rs. 1.13 crores for the development of brackishwater shrimp culture in the state in its 1993-94 budget. The phase of transformation in shrimp culture in the state is accelerated through planned infrastructure development provided by MPEDA, a nodal agency in the promotion and expansion of brackishwater aquaculture of shrimp. With the organisational support given by the State Government, the financial assistance extended by nationalised commercial banks and with the availability of brackishwater potential, techno-economic viability, favourable environment, and enterprising farmers, Andhra Pradesh is all set to bring in 'Shrimp Revolution' in the country.

Aquaculture of shrimp under brackishwater environment, apart from providing subsidiary income and employment to farmers and labourers, also contributes to augment the foreign exchange earnings and to accumulate foreign exchange reserves.

---

15. Subramanyam, M.V., "Towards greener pastures-From Agriculture to shrimp farming", *Frontline*, Vol. II (17), Aug, 1994, pp. 13-26,
16. Staff Report, "Chepala Pempakaniki 6 Kottlu", *Udayam*, 3rd March, 1993, p. 5.

These advantages, therefore, make shrimp farming an export oriented, priority allied activity in the process of India's developmental planning.

The potential, development and importance of fisheries, with special emphasis on brackishwater shrimp aquaculture however is analysed in detail in Chapter 2.

Present Study

Agriculture sector by itself cannot solve all the socio-economic problems responsible for unemployment, disguised unemployment, under employment and poverty in the rural areas. In the absence of increase in the labour absorption capacity of the industrial sector, rural employment position seems to be bleak. It is a recent phenomenon that allied activities in rural areas, along with agriculture are seen as potential sources for providing additional employment and income. Apart from allied activities like dairy, poultry, sericulture, horticulture and inland fish culture, brackishwater shrimp culture is considered as a promising activity to provide employment and subsidiary income to the rural poor and to augment the foreign exchange earnings to the government.

The research done in the past by agricultural economists has almost ignored the vast area of its allied activities. This lacuna has been pointed out time and again by academic and expert bodies. The research on fisheries economics in general and economics of brackishwater shrimp culture in particular was rather scanty. Though a few studies are available on economics of fisheries, they are only on inland fish culture, and are not comprehensive in nature. The recent study of Hariprakash on 'Economics of inland fish culture : a case in West Godavari District of Andhra Pradesh' is somewhat comprehensive in nature, but it is on inland fish culture only. So far there is no comprehensive study on brackishwater shrimp culture with special reference to the state of Andhra Pradesh which covers economic aspects like costs and returns, employment, marketing structure etc. Understanding of these issues is very essential in order to recommend correct policy measures to promote the growth of brackishwater shrimp culture, which is considered as the 'Extreme Focus Area' from the stand point of India's export promotion and foreign exchange earnings. Micro level studies in this regard at this stage of the industry certainly facilitate to draw definite conclusions for sorting out relevant problems. The present study endeavours to fulfil this requirement.

Methodology

Commensurate with the objectives of the present study, Krishna District of Andhra Pradesh is purposively selected mainly because of the availability of vast brackishwater area to the extent of about 24,662 hectares. Out of 17,000 hectares of brackishwater area found suitable for immediate shrimp farming in the state, 6,239 hectares of area is located in Krishna District alone. The Tiger Prawn (penaeous monodon) which is having good export market is the chief cultured item in this district. The 8,727 sq. km. area that the district has, about 800 sq. km. lie hugging the coast line. This area has natural features suitable for the development of brackishwater shrimp farming. The brackishwater shrimp culture activity has been in operation in this district for more than a decade. But the scientific a aquaculture of shrimp commenced only in the early nineties. Out of 37 brackishwater Fish Farmers Development Agencies established in the country to promote brackishwater aquaculture, one is at Machilipatnam, the district headquarters. There is also one shrimp farming unit and a field office of MPEDA at Vijayawada, which is an important commercial town of this district. In addition, Krishna District is one of the four districts selected in the entire country to start the World Bank assisted Shrimp and Fish Culture Project with an estimated capital outlay of about Rs. 83.13 crores. In view of all the above mentioned factors, Krishna District has been selected for this study.

Selection of Mandals

Krishna District is divided into 50 revenue mandals, seven out of which are located along the coast line. Out of these seven mandals shrimp culture is in operation in six mandals. Out of these six mandals, three mandals viz., Bandar (Machilipatnam), Bantumilli and Kruthivennu mandals have been chosen for the study as more than fifty per cent of the brackishwater spread area in the district falls within these three mandals. From among these mandals one village from each mandal has been selected. They are Tadivennu in Kruthivennu mandal, Manimeswaram in Bantumilli and Kona in Bandar mandal. The selection of these villages is

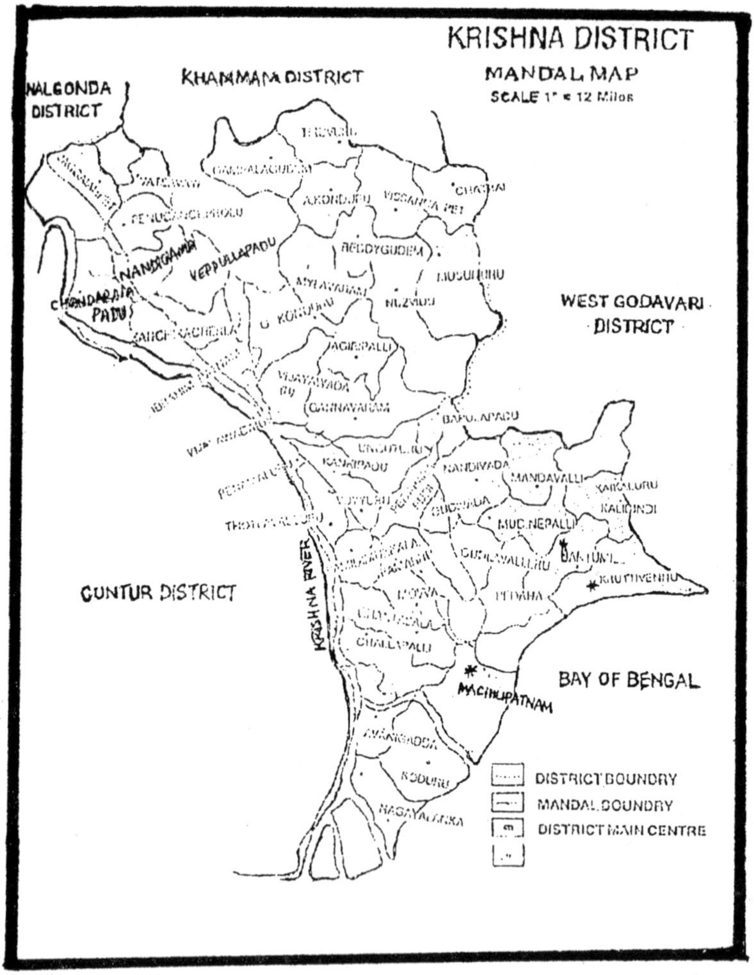
KRISHNA DISTRICT
MANDAL MAP
SCALE 1" = 12 Miles
NALGONDA DISTRICT
KHAMMAM DISTRICT
WEST GODAVARI DISTRICT
GUNTUR DISTRICT
BAY OF BENGAL
KRISHNA RIVER
NANDIGAMA
VEPPULLAPADU
REDDYGUDEM
MUSUNURU
NUZVIDU
GANNAVARAM
BAPULAPADU
KANKIPADU
NANDIVADA
MANDAVALLI
KAIKALURU
KALIDINDI
MUDINEPALLI
BANTUMILLI
KRUTIVENNU
PEDANA
CHALLAPALLI
MACHILIPATNAM
AVANIGADDA
KODURU
NAGAYALANKA
DISTRICT BOUNDRY
MANDAL BOUNDRY
DISTRICT MAIN CENTRE

Fig 1.1 :

based on the strength of the farmers who have experience in shrimp farming and also their representation from various size groups.

Selection of Farmers

The three selected villages altogether have 430 shrimp farmers belonging to various size groups whose holdings range from about 0.50 acres to 35 acres. The farmers are classified into four size groups namely marginal (below 2 acres), small (2-5 acres), medium (5-10 acres) and big (above 10 acres). Out of these 430 farmers belonging to various size groups, 215 farmers *i.e.*, 50 per cent are selected for the study. The number of farmers selected in each size group is based on the probability proportion to the number of such farmers available. Finally random sampling technique is employed to select the individual farmers from each size group.

Table 1.1. : District-Wise Distribution of Brackishwater Area in Andhra Pradesh

District	*Brackishwater Area (Hectares)*	*Brackishwater Area for immediate development (Hects.)*
Srikakulam	3,932.42	1,316.20
Vizianagaram	12.00	10.00
Visakhapatnam	1,291.70	720.20
East Godavari	24,837.40	3,353.00
West Godavari	3,467.50	1,532.00
Krishna	24,361.99	6,239.00
Guntur	1,594.80	1,376.00
Prakasam	1,411.53	619.00
Nellore	2,753.00	1,841.00
Total	63,662.34	17,006.40

Source : 'Culture Fishery Development in Krishna District of Andhra Pradesh : A Case Study of Krishna District by J.V.H. Dixitulu, for Agro-Climatic Regional Planning Unit, Planning Commission, India, Sardar Patel Institute of Economic and Social Research, Ahmedabad, 1990, pp. 3-4.

Data Base

The required data was collected from primary and secondary sources. The secondary data and information required to examine the growth and status of brackishwater shrimp culture in India and in Andhra Pradesh in particular were collected from proceedings of the workshops conducted, course manual prepared by Indian Council of Agricultural Research (ICAR), Central Institute of Fisheries Education (CIFE), Central Marine Fisheries Research Institute (CMFRI) on brackishwater shrimp farming and the manuals and reports prepared by South East Asian Fisheries Development Centre (SEAFDC), Association of South East Nations (ASEAN), Manila, Philippines, the fishery statistics provided by Commissionerate of Fisheries, Government of Andhra Pradesh, Regional Offices of Marine Products Export Development Authority of Visakhapatnam, Vijayawada and from the publications of MPEDA.

Table 1.2 : Distribution of The Total Shrimp Farmers In the Selected Three Villages

S. No.	Farmers	*Manimeswaram*		*Tadivennu*		*Kona*		*Total Three Villages*	
		Number of Farmers	*Percentage to total*	*Number of Farmers*	*Percentage to total*	*Number of Farmers*	*Percentage to total*	*Number of Farmers*	*Percentage to total*
1.	Marginal Farmers	84	58.33	40	28.57	92	49.31	196	45.58
2.	Small Farmers	28	19.44	96	68.57	40	27.39	164	38.15
3.	Medium Farmers	20	13.88	2	1.43	18	12.33	40	9.30
4.	Big Farmers	12	8.35	2	1.43	16	10.97	30	6.97
5.	All	144	100.00	140	100.00	146	100.00	430	100.00

Table 1.3 : Selection of Sample Farmers from the Sample Villages

S. No.	*Farmers Size Group*	Manimeswaram		*Tadivennu*		*Kona*		*Three Villages*	
		Total	*Sample*	*Total*	*Sample*	*Total*	*Sample*	*Total*	*Sample*
1.	Marginal Farmers	84	42	40	20	72	36	196	98
2.	Small Farmers	28	14	96	48	40	20	164	82
3.	Medium Farmers	20	10	2	1	18	9	40	20
4.	Big Farmers	12	6	2	1	16	8	30	15
5.	All Farmers	144	72	140	70	146	73	430	215

Primary data were collected through survey method with the help of specially designed schedule containing a comprehensive list of items of information on brackishwater shrimp culture. Some of the items included in the schedules are : various types of costs incurred, returns obtained from sales and marketing structure. The sample farmers were interviewed to elicit the required information, regarding costs, prices, employment opportunities and marketing.

The field data obtained from the sample shrimp farmers has been supplemented by the information furnished by Brackishwater Fish Farmers Development Agency (BFFDA), Machilipatnam Office, MPEDA, Vijayawada, investigators from Central Institute of Brackishwater Aquaculture (CIBA), Madras, and was used for cross-checking of the information given by the sample shrimp farmers. Besides this, discussions were also held on various issues related to shrimp farming with Assistant Director of Fisheries, Krishna District, office bearers of Shrimp Farmers Co-operative Societies, and with some knowledgeable persons in shrimp culture in Krishna District.

Techniques of Analysis

Tabular analysis with percentages are used for most of the analysis. Simple ratio methods like net returns over investments, output-input ratio, are used to examine the efficiency of shrimp farming. Multiple regression analysis is used to estimate the production function *i.e.,* to analyse the factors affecting the cultured shrimp production.

Reference Period

The study refers to one production year i.e., 1993-94. In shrimp farming, the farmers generally raise two crops in a year with four and half to five months duration for each crop. Therefore, the costs and returns are calculated for two crops and used for the purpose of estimating the net returns. 1993-94 is also considered as the year of shrimp.

Scheme of the Study

The study is organised into eight chapters. Chapter 1 comprises introduction to the problem, and methodology of study.

Chapter-2 deals with the significance of fisheries, and shrimp in particular in India. Chapter - 3 encompasses the genesis and evolution of aquaculture. It also deals with the status of brackishwater shrimp culture in Andhra Pradesh. Chapter-4 presents a review of earlier studies and specifies the problem for the present study. In Chapter-5 a profile of Krishna District, the study area and the socio-economic characteristics of sample brackishwater shrimp farmers are covered. Costs and returns in shrimp culture for different size groups of farmers is analysed in Chapter-6. Chapter-7 portrays employment generation and marketing structure of brackishwater cultured shrimp in the study area. The last chapter presents, the summary and conclusions of the study.

Chapter 2

Fisheries and the Brackishwater Shrimp Culture in the National Economy

Introduction

This chapter deals with the prominence of fisheries and fish potential in general and brackishwater shrimp culture in particular in Indian economy.

The knowledge of occurrence of fish in India dates back to 3,000 B.C. Fish remains with cut marks, which indicate their use as food, were obtained from excavations of Mohenjodaro and Harappa of the Indus Valley Civilisation (2500 to 1500 B.C.).[1] Fishing is an age-old caste based activity in India. Fisheries (fish and shellfish) occupy a prominent place and play a phenomenal role to-day, in the Indian national economy. As an integral part of agriculture and as an allied activity, fisheries occupy a significant place particularly in the economic life of the rural poor in India. The development of fisheries and allied activities have a special importance in India because, (i) this sector is eminently suitable to assist a large section of backward and economically weaker sections of rural community, (ii) it provides a high subsidiary food at reasonable prices with a high quality protein and (iii) it has a large export potential.

At its present state of economic development, India relies heavily on the exploitation of its natural resources, and fisheries is one among them. The contribution of fisheries, and their importance

1. Sarma, B.C., "Social Issues in Aquaculture", in K.K. Trivedi (Ed.) *Aquaculture Research Needs for* 2000 AD., Oxford & IBH Publishing Ltd., New Delhi, 1988, p. 399.

in the Indian economy can be examined under two headings : (1) Fisheries as a source of food and (2) as a source of income and employment.

Fisheries As A Source of Food

The development of fisheries in India is of great importance, especially, from the stand point of nutrition and achievement of self-sufficiency in food.

Malnutrition is a world-wide problem and India is no exception. It is estimated that about 10 million people die every year in the world because of starvation and malnutrition.[2] The problem of protein is much more acute in a populous country like India with chronic food shortage and malnutrition and the diet of a large number of masses constitutes cereals and other inferior type of food.[3] In this context the contribution made by fishery resources to human nutritional needs is significant and may represent the only readily available protein rich food for the people in a developing country like India.[4] Nutrition experts agree that fish with the addition of a variety of vegetable products constitute a completely balanced diet.[5] From the point of view of human nutrition, fish food is not only easily digestable but is also rich in Aminoacids like fatty acid namely 'EICOSAPENTAENOIC' acid if fish is known to reduce the cholestrol level on blood and save more human beings from coronary disorders. Vitamins such as A,B,C,D, B-Complex, B12 and minerals like calcium, phosphorus, iron sodium potassium, magnesium and sulphur are also present in good quantities in fish.[6] Fish in many ways is even more nutritious than most meats of the warm blooded animals.[7] The protein content of fish varies from 17 per cent to 20 per cent of

2. Santhanam, E., Sukumaran, N., and Natarajan, P., A Manual of Fresh Water Aquaculture, Oxford & IBH., New Delhi, 1990, p. 1.
3. Gopalan, C., Rama Sastri, B.V., and Balasubramanian, S.C., *Nutrition Value of Indian Foods*, National Institute of Nutrition, Indian Council of Social Research, Hyderabad, 1976. p. 4.
4. David James, "The future for the fish in nutrition", *Inofish Marketing Digest*, No. 4/84. Inofish International, Kualalumpur, Malaysia, p. 41.
5. Gopalan, C., et al., Nutrition Values of Indian Foods. *Op. Cit., Pp 40*
6. Santhanam, R., et al., A Manual of freshwater aquaculture, *Op. Cit.*, p. 2.
7. Food and Agriculture Organisation, Fish Culture in Central East Africa, Rome, 1955, p. 3.

dressed weight. Fish flesh constitutes about 11 per cent of animal protein now consumed by people.[8] As protein from live stock becomes more and more costly, the relatively inexpensive fish protein forms the main source of animal protein in poorer communities throughout the world. In many tropical countries, fish consumption now exceeds that of all other animal protein. The average per capita consumption of fish currently exceeds thirteen kilograms a year. In South East Asia, about a million people relay predominently on fish for animal protein to-day.[9] The development of under exploited and unexploited fisheries resources, both capture and culturable, offer a promising remedy to the problem of protein food shortage. The exploitation of fishery resources would materially raise the standard of the peoples' diet, and constitute an important source of food next to agriculture and animal husbandry, if the resources of fisheries are systematically and scientifically exploited.[10]

Fisheries as a Source of Income and Employment

Fishery is an important sector in most of the developed and developing countries of the world from the stand point of income and employment generation. The experience in these countries indicates that the growth of the fishery sector stimulates the development and employment in related industries which contribute significantly to the total economic growth of the country. These industries include transport, shipbuilding, repairs, processing of fish, ice making, cold storage, freezing facilities and animal food production etc. In many cases it contributes over 5 per cent to the gross national product. Some 20 million fishermen are found to be engaged in fishing in these countries and atleast an equal number are employed in related industries.[11]

In India fisheries sector contributes about 2 per cent to agriculture as its sub-sector and less than 1 per cent to the GDP.

8. Jackson Roy, I., "Fisheries and Future World Food Supply", World Fisheries *Policy-Multi-Disciplinary Views*, (Ed.), Brain J. Rothschild, University of Washington Press, Seattle and London, 1972, p. 11.
9. Younis, A.S., and G. Donaldson, *Fishery : Sector Policy Paper*, The World Bank, Washington, 1982, p. 20.
10. Subba Rao, N., *Mechanisation and Marine Fishermen : A Case Study of Visakhapatnam*, Northern Book Centre, Delhi, 1988, p. 2.
11. Younis, A.S., and G. Donaldson, *Fishery Sector Policy Paper*, *Op. Cit.*, p. 5.

However, fisheries in India is important as it provides employment to an estimated 9 million people, mainly from the poor coastal communities. It contributes in raising nutritional standards and is a foreign-exchange earner. Moreover, the fishery sector has been growing rapidly at an annual average rate of 6 per cent over the past one decade with high potential for future growth, particularly in domestic market where the demand for fish is expected to be doubled by the year 2000 AD. Between1980 and 1990, fisheries production increased from a total of 2.5 million tonnes to 3.7 million tonnes of which 60 per cent came from marine capture fisheries. By 1990, exports of marine products increased to about $360 millions which has double that of 1980, with shrimp exports accounting for about 90 per cent. India occupies seventh position among the fish producing countries and earns foreign exchange of about Rs. 2,503 crores through marine products exports in 1994.[12] However, for India to meet its growing domestic needs and to maintain its position in the export market, new initiatives are necessary, requiring an expansion and diversion of the conventional fishing areas and bring into operation more offshore and deep sea areas in marine fisheries, and development of freshwater and brackishwater aquaculture.[13]

Fisheries' Potential in India

The aquatic resources of fisheries are varied and extensive. However, various sources of fisheries can be broadly divided into two types, viz. (1) Marine sources of fisheries and (2) Inland sources of fisheries.

Marine Fisheries

India has excellent fishing area in marine fisheries. It includes the two wide arms of Indian Ocean, *i.e.*, the Bay of Bengal and Arabian sea. There are smaller gulfs, bays, and creeks along the coast. In addition mangrove swamps and broad coral reefs form the mainstay and provide extensive marine fishing grounds in India.[14]

12. Varghees P.U., "Aquaculture Production to augment export of marine products from India", *Indaqua-Aquaculture Exposition, Op. Cit.*, Jan. 27-29, Madras, 1995.

13. India : Brackishwater and Inland Fisheries Project, *FAO/World Bank Cooperative Programme, Staff Appraisal Report*, Report No. 70/91, CP-IMD 85, Jun., 12, 1991, pp. 1-2.

14. Chopra, B.N., (Ed.), *Hand Book of Indian Fisheries*, Government of India, Ministry of Agriculture, Agricultural Series, NO. 44, New Delhi, 1951, p. 8.

India has a coastline of 6,500 KM. (excluding that of Andamans, Nicobars and Lakshadweep) with a continental shelf of 5.8 million sq. kilometers into which a number of large and perennial rivers discharge their silt-laden waters. Sea fishing has been an occupation of coastal people in India from times immemorial , forming an integral part of the country's maritime heritage.[15] Among the maritime nations that border the Indian ocean, India is the largest and the most powerful fishing country. It is estimated that Indian ocean can yield a fish potential of 14.39 million tonnes per annum, of which only one-fifth is exploited (2.88 million tonnes). Even in that exploited 1/5, the India's catch in Indian ocean is only meagre 1.2 million tonnes.

The Exclusive Economic Zone (EEZ) of India which has been extended upto 200 nautical miles can yield a fish potential of 4.47 million tonnes per annum. Of this as much as 2.26 million tonnes or 50.6 per cent is in the inshore zone, 1.71 million tonnes or 38.3 per cent is in the deep sea zone and only 0.50 million tonnes or 11.2 per cent is in the depth beyond 200 metres which incidentally accounts for 79.4 per cent of the total area of the EEZ.[16]

The important and widely available species of fish caught in Indian ocean are shrimp, oil sardine, mackerel, seer, elasmobranches, catfish, ribbon fish, perches, etc.

Inland Fisheries

The extensive backwaters, tidal estuaries, lagoons and swamps, scattered along the coast line and also a large number of rivers, streams and channels and an equal number of perennial and semi-perennial lakes, reservoirs, tanks, ponds and other stretches of water are a rich potential sources of inland fisheries.[17]

Inland fisheries in India consist of three types of activities, viz., (a) Capture fisheries (mainly riverine), (b) stocking-cum-

---

15. George, P.C., "Fisheries Planning and Development in India", *Proceedings of the Indo-Pacific Fisheries Council*, 16th Session, Section III, Bangkok, 1974, p. 154.

16. George, P.C., Anthony Raja, B.T., and George, K.C., "Fishery Resources of the Indian Economic Zone", in *Souvenir* issued on the occasion of the Silver Jubilee Celebrations of the Integrated Fisheries Project, Cochin, Oct., 1977.

17. Chopra, B.N. (Ed.), Hand Book of Indian Fisheries, *Op. Cit.*, p. 8.

capture fisheries (Regervoirs, Lakes etc.), (c) Culture fisheries (ponds, tanks), in fresh and brackishwaters.[18] Of the 1.6 million hectares of culturable fresh waters in the form of ponds and tanks of the country, only about 0.6 million hectares are at present under utilisation and out of the estimated brackishwater area of about 12 lakh hectares, only 65,000 hectares are mainly under shrimp farming.[19] The other sources of fisheries include 2.9 million 'hectares of reservoirs and more than 27,000 kilometers of rivers with their main tributaries, are only partially exploited. In addition to this,there is a wide net work of irrigation canals and channels running through an area of over more than 1,10,000 kilometers.[20] With the evolution of new culturable technology at various centres, it is estimated that the average annual harvest from fish farms has increased from 600 Kgs. per hectare a few years ago to around 9000 Kgs. per hectare.[21] The highest production of more than 10,000 Kgs. per hectare per crop has been achieved by some progressive entrepreneurs in Tuticorin of Tamil Nadu by adopting intensive method of farming. The productivity level under scientific extensive method ranges from 500 top 2000 Kgs. per hectare.[22]

The important varieties of fish that are caught in the inland area are Cutla, Rohu, Carps, Catfish, Feather backs, Murrels, Hilsa and the all important shrimp.

Brackishwater Shrimp Culture in India

Brackishwater fish farming as an empirical practice has been known to India from times immemorial.[23] The 'Filtration' or

18. Subrahmanyam, N.S., and Anjaneyulu, P.S.R., "Inland Fisheries in Andhra Pradesh : Present Status and Prospects of Development", *Fishing Chimes*, Dec., 1992, p. 49.
19. MPEDA, *Hand Book on Shrimp Farming*, *Op. Cit.*, p. 6.
20. Bunder N. Sud, "Slow Development of Inland Fisheries, *The Indian Express*, Feb., 21, 1974, p. 6.
21. Jhingran, V.G., "Scope and Role of Inland Aquaculture in India's Economy", *Indian Farming*, Vol. 25 (6), Sep., 1975, p. 11.
22. Jhingran, V.H., "Scope and Role of Inland Aquaculture in India's Economy", *Op. Cit.*, p. 7.
23. Silas E.C., Vedavyasa Rao, P., "Fish and Shellfish Production through aquaculture in Brackishwaters of India", in U.K. Srivastava, B.H. Dholoki (Eds.), *Brackishwater Aquaculture Development in India*, Concept Publishers, Delhi, 1985, p. 119.

'Trapping' method of brackishwater shrimp farming in the inter tidal zone of coastal belt of India especially in West Bengal (Bheries) and Kerala (Pokkali fields) has been a very old practice.[24]

The (Khar) lands of Karnataka and the (Ghazani) fields of goa are more recent conversions to this traditional trapping system of shrimp farming. In this system the shallow brackishwater impoundments are auto stocked with natural seed through high does and are being harvested by filtration by fixing suitable screens in the sluices.[25] Though nearly two-thirds of the brackishwater area under culture is still under traditional practice, shrimp farming in India can be seen to have undergone a tremendous change from the traditional method of culture to advanced scientific farming.

There are more than 50 varieties of shirmp and six of them are identified for brackishwater aquaculture. Among these six varieties panaeid shrimp, the 'Black Tiger prawn' which is popularly known as Jumbo Tiger Prawn (P. Monodon) is the most popular species for culture because of its high growth potential and price.[26] It attains maximum size of 330 mm. in natural habitat, 240 mm. in pond growing conditions. Tiger prawn is highly relished by gourmets all over the world. The second important species is the 'Indian White Shrimp' (P. Indicus) which is versatile but has less growth potential than P. Monodoan. Its size ranges from 140 to 170 mm. These two species of shrimp are cultured extensively and almost exclusively under brackishwater aquaculture in India.

There is a world wide shift in favour of modern scientific shrimp farming in recent times. However, shrimp farming in an organised way attracted the attention of the government and the people living in coastal areas in India only as late as in the eighties. Various organisations have initiated efforts to develop appropriate technologies for shirmp farming, under brackishwater environment, because it is considered as money spinner with tremendous export potential and is also found to be the best

24. Balakrishna Nair, N., "Aquaculture in India-Problems and Prospects", *Aquaculture Productivity, Op. Cit.*, p. 63.

25. Srivastava, U.K., Amarjeet Singh and Martin Santhankumar, J., "Economics of Brackishwater Culture in India-Some Preliminary Evidence", in U.K. Srivastava and M. Darma Reddy (Eds.), *Fisheries Development in India,* Concept Publishing Co., New Delhi, 1983, p. 398.

26. MPEDA, *Hand Book on Aqua Farming : Shrimp, Lobster,* Mucrab, MPEDA HOUSE, KOCHI, 1992, p. 11.

alternative for the convential sources of shrimp to step up the shrimp production. Therefore, brackishwater shrimp farming has taken roots and has become an attractive enterprise in India.

In view of the vast potential offered by this sector, the pace of development has also been fairly rapid over the last few years in shrimp industry. Among the marine products exported from India shirmp is the dominant item. About 650 million dollars worth transactions takes place every year all over the world in this discipline of which India's share is around 20 per cent.[27] It accounts for 70 per cent in terms of value and 40 per cent in volume in our sea food exports.[28] Shrimp is the real bread earner of the export industry and a dependable source of employment and livelihood for several thousands of people engaged in its culture particularly in the rural areas.

The concept of 'satellite shrimp farming' [29] is also coming up in all maritime states. A few private companies have already put up 100 per cent export oriented units (EOU) in shrimp production and exports. With the technical improvements made in shrimp aquaculture, the interest shown and initiative taken by the farmers in this activity, and the support provided by the Government agencies to this venture, India will one day certainly be 'The Kingdom of Aquaculture' according to Sakthivel. [30]

Significance of Brackishwater Shrimp Culture in Indian Economy

At present, the faster rate of growth of population of our country places a very heavy demand on its natural resources such as land, water, forests, mines etc. Already existing problems like malnutrition, unemployment, rural backwardness, depleting foreign exchange reserves etc., are being aggravated further due to

27. Staff Report, "A. P. Golmine of Prawnculture", *Deccan Chronicle*, Jul. 15, 1993, p. 16.

28. Kannan, R., "Aquaculture gains importance", *The Hindu-Survey of Indian Industry*, 1993, p. 401.

29. Under Satellite farming system the principal farmer (corporate farm) play a major role. It sets up nucleus' i.e. its ownfarm hatchery, feed mill, R & D Unit,processing plant not only for its own needs but also to cater to the needs of other small shrimp farmers (See : *Hand Book on shrimp farming*) MPEDA, 1992, p. 4.

30. Sakthivel, M., "Indian Aqua industry has immense growth potential", *Aqua-International*, Hyderabad, Nov., 1993, p. 37.

the mounting population pressure. Aquaculture of shrimp in particular is expected to play a phenominal role in Indian Economy wherein much of its brackishwater resource potential is either under exploited or unexploited. The importance of brackishwater shrimp culture is also greatly felt,as that of agriculture and animal husbandary, because it provides great opportunity for production of valuable and healthy fish food for people, employment and income to the rural poor, foreign exchange earnings to the exchequer, better land and water use and overall development of the country. Aquaculture of shrimp in this context has been considered as a 'Sunrise Industry'.[31] The importance of brackishwater shrimp culture can be analysed under three separate headings : *i.e.*, (1) as a source of protein rich food, (2) as a source of employment for rural poor and (3) as a source of foreign exchange earnings.

Shrimp as a Source of a Protein Rich Food

In the context of increasing health consciousness in the modern world fish and fish products are considered to be among the safest foods of animal origin, and India is no exception to this world-wide trend. Due to an increasing shift towards high protein, low cholesterol food habits, shrimp finds growing acceptance all over the world. The problem of protein gap is much more acute in a developing country like India which is experiencing high population pressure. Since the food available through agriculture and animal husbandary is insufficient to feed the teaming millions of the country, the diet of a large number of people is unbalanced in India. Therefore, there arose a need to seek other sources of staple food to supplement the present diet.[32] Shrimp, which is

31. Sakthivel, M., "The problems confronting the development of aquaculture in India and possible remidies", *Aquaculture Productivity, Op. Cit.*, p. 747.

32. In India nearly 326 million people are subjected to malnutrition, according to the survey conducted by the Indian Institute of Public Opinion. Unless the protein intake is increased it is feared that about 50 million children in India will die of malnutition and another 82 million suffer from retarded growth. While measures to increase the production of protein from land, directly through crops and indirectly through livestock, are far from adequate to tackle chronic wide-spread malnutrition problem in the country (See : Indian Institute of Public Opinion, Nutrition and Health'); The condition of productivity-A framework for a policy for the weak (1970-80). This leads to the conclusion that India must also explore other possibilities of obtaining food. In this context fisheries including shrimp provide immense scope to fill the gap of food shortage.

also an aquatic organism is realised as protein rich food with universal appeal. Desirability of shrimp as food item is also partly due to the absence of bones in them.[33]

The domestic consumption of shrimp in India has increased from about 68,700 tonnes per year in 1967 to 1,13,800 tonnes in 1987. This is at an average growth of 2.6 per cent. However on per capita basis, the average annual shrimp consumption has remained relatively low (less than 150 grams) throughout the past two decades. It is observed, that this is partly due to the fact that a large number of Indian are vegetarians and mostly due to the high price of shrimp in relation to the purchasing power of the majority of the population. In addition, the domestic market could not consistently supply quality shrimp and frequently received only second rate shrimp.[34]

With the increase in brackishwater area under shrimp culture a reasonable decline in the price and consistant supply to the domestic market, the per capita consumption of shrimp of people of India and their standard of diet would certainly raise.

Shrimp Culture as a Source of Employment to Rural People

The development of brackishwater shrimp aquaculture is of great economic significance from the stand point of rural employment in India. Development of rural areas is possible only through the generation of adequate gainful employment. Several observations made by aquaculturists and economists have conclude that brackishwater shrimp culture activity is economically viable as a supplementary or as an alternative source of employment and income.

Shrimp farming development under brackishwater system holds immense employment potential for the local people. It provides direct employment to at least 2 persons for managing and operating one hectare culture area. It also provides indirect employment to the extent of 1100 man days per hectare, during construction, input supplies, handling of harvest, marketing etc., In other words if an area of 1 lakh hectares is developed fro shrimp

33. Yap, W.G., Apud. F.D. and Primavera, J.H., *Manual on Prawn Farming*, Aquaculture Extension Manual No. 5, Manila, July, 1979, p. 1.

34. India : Brakishwater Shrimp and Fish Culture Project, *Staff Appraisal Report, Op. Cit.*, p. 68.

farming, it will provide direct employment to 2 lakh people and generate indirect employment to another 3 to 5 lakh people in allied activities such as hatchery operation, feed production, processing, marketing etc.[35]

The total manpower requirement category-wise for the phased development of brackishwater shrimp culture in India upto 2000 A.D. is presented below.

Category of Personnel

Year	*Fishermen*	*Operators*	*Graduates*	*Postgraduates*
1985	23,000	40,000	1,400	400
1990	4,50,000	80,000	23,000	2,000
1995	12,00,000	2,00,000	67,000	20,000
2000	17,00,000	3,00,000	1,00,000	30,000

The details show that there is going to be an increasing trend in the manpower requirement of brackishwater aquaculture of all categories of workers. By 2000 A.D. India would be able to employ 17 lakh fishermen, 3 lakh operators, 1 lakh graduates and 30,000 postgraduates from different disciplines.[36]

Seed is an important input for aquaculture. As the shrimp seed through hatcheries is unable to meet the growing demand, a large number of farmers are depending on wild seed. India is endowed with rich natural seed resources of shrimp in the vast stretch of coastal lands for farming. Several thousands of rural families are engaged in the collection of shrimp seed to be sold at an attractive price of Rs. 500-1000 per thousand, depending on the demand supply situation. About 50,000 families in the Sundarbans (a group of islands covered with the mangrove fore sets at the confluence of the Ganges and Bay of Bengal) in West Bengal are engaged in the collection of shrimp seed.[37] In East Godavari, West Godavari, West Godavari and Krishna Districts of Andhra Pradesh also several thousands of people of all age groups and sex are able to live on the seed collection and sale in the coastal villages.[38]

35. Mpeda, *Hand Book on Shrimp Farming, Op. Cit.*, p. 21.
36. Divedi, S.N., "Prawn culture : Challenges and Possibilities", in U.K. Srivastava, et al. (Eds.), *Brackishwater Aquaculture Development in India, Op. Cit.*, pp. 111-117.
37. Tapash Ganguly, "In Deep Waters", *The Week*, Jan. 9, Vol. 12 (3), 1994, pp. 44-46.
38. Staff Report, Vithanala Dash alone Veladhimandiki Upadhi", *Udayam*, Nov. 20, 1993, p. 12.

Labour who used to find it difficult to earn Rs. 10 per day are able to earn anywhere between Rs. 150-200 per day by way of collecting shrimp seed.[39]

Several educated youth are also involved in the shrimp culture, having realised that this enterprise is economically rewarding and intellectually satisfying. According to Pillay, properly planned and appropriately integrated aquaculture could certainly increase the output, income and employment and improve the living standards of the rural poor in India.[40]

Shrimp Culture as a Source of Foreign Exchange

Growing deficit in the balance of payments is a problem of concern to the exchequer in India in recent years. Though the government has managed to overcome the balance of payments crisis of mid 1991 and averted the dreaded defaulting stage and built up substantial foreign exchange reserves, the position is not yet totally out of danger. The external debt burden has been increasing at a beweildering rate, and at present it stands at a staggering figure of about 90 billion dollars and is expected to increase further and cross $ 100 billion mark by the turn of the century. At this time of mounting external debt, debt servicing charges and increasing deficit in the balance of payment, boosting of exports through export promotion and diversification to augment our foreign exchange reserves seems to be the desirable option to India. In this context, aquaculture of shrimp assumes greater significance as a source of foreign exchange.

Among the marine products exported from India, shrimp is considered as the dominant item by virtue of its unique taste, delicacy in diet and universal acceptance. Being the fastest moving commodity in the global market it is foundly referred to as the 'darling', the 'Pink Gold' of the sea food industry.[41]

The major impetus to marine fisheries development in India has come from the export potential of shrimp since 3rd August, 1953 when the first consignment of Indian shrimp was exported

39. Staff Report : *Eenadu*, June 24, 1993, p. 1.

40. Pillay, T.V.R., *Planning of Aquaculture Development-An Introductory Guide*, Fishing News Book, England, 1977, p. 30.

41. Balakrishna Nair, N., "Aquaculture in India-Problems and Prospects", *Aquaculture Productivity, Op. Cit.*, p. 62.

from Cochin Port to United States. However, the catch of wild shrimp from natural sources is said to be stagnant for some time due to over exploitation, while the demand especially in the principal markets of U.S., Japan and Western Europe have been on the rise. Therefore, the efforts are concentrated on shrimp farming in India in recent times to augment the production. The lucrative little shrimp thus became the focal point of an all time enterpreneural wave in India and very much a part of the booming billion dollar aquaculture industry of the world.

According to Rosario G. Manalso, shrimp culture has attracted considerable attention in recent years not only because of its value as food supply but also because of its high potential as foreign exchange earner.[42] The thrust in culture production of shrimp is mainly due to the fact that it is considered as a foreign exchange spinner.

The value of marine exports which was 24 crores in fifties, increased to Rs. 1767.43 crores by 1992-93. Of this, shrimp item alone accounted for 1180.26 crores. It is anticipated that shrimp exports will go up by an annual level of Rs. 3,400 crores by the end of the 8th Five Year Plan according to the conservative estimate of Sakthivel (1993). In his opinion aquaculture industry of India should be in a position to earn at least 3000 million U.S. dollars per year by way of shrimp exports. The value of shrimp exported from India during 1994 has been estimated at Rs. 2,503 crores. The scope for future growth of brackishwater shrimp culture is immense in India. It has not even touched the fringe of its brackishwater resources for shrimp farming. It can still bring in more area under shrimp culture and produce more shrimp because selling shrimp is not a problem for India. Indian farmed shrimp is becoming a major attraction to many overseas buyers, including the chief executive of leading chain stores.[43]

India presently is considered as the world leader in shrimp production and export. With very high up beat global market conditions, the shrimp farming has been identified by the Ministry of Commerce, Government of India as 'thrust area' and brought it under extreme focus category for export promotion.

---

42. Rosario G. Manalso, *Manual on Pond Culture of Penaeid Shrimp*, Manila, Philippines, 1978.
43. Sakthivel, M., "Indian Aqua Industry has immense growth potential", *Aqua International*, *Op. Cit.*, pp. 36-38.

Chapter 3

Aquaculture : Genesis, Evolution and Brackishwater Shrimp Culture in Andhra Pradesh

Introduction

Aquaculture is the farming and husbandry of economically important aquatic organisms under controlled conditions. It includes all aspects of production of aquatic organism in captivity comprising either some or all stages of their life and their food and the resultant of marketable products in the habitat of fresh, brackish and sea waters.[1]

The term 'aquaculture' encompasses a number of activities like traditional extensive fisheries, management in large bodies of freshwater ponds such as lakes and reservoirs, semi-intensive and intensive culture in small tanks and ponds and brackishwater areas, and finally sea ranching and mariculture. Aquaculture which was originated as a subsistance level activity in very ancient times has turned into a booming industry at present. The impressive growth has also been stimulated and supported by modern scientific research. It is believed that the current world production of 5 million tonnes of fish may increase by seven to ten-fold by the end of this century, mainly due to the increase in aquaculture practices.

In this chapter an attempt is made to present the genesis, evolution of aquaculture with special emphasis on the brackishwater shrimp culture in Andhra Pradesh.

1. Ministry of Agriculture and Irrigation, Report of the National Commission on Agriculture, Part VIII, Fisheries, Government of India, New Delhi, 1976, p. 22.

Genesis of Aquaculture

Primitive man, especially in the Pleistocene, was probably a vegetarian, since he did not have any implements or gear to catch fish. It is only in the Neolithic age and the subsequent period in which prehistoric man could make better stone tools so that fishing became known to him. At that time the way of life was hunting, fishing and gathering of wild crops. Between 7000 and 5000 B.C. hunters and fishermen were spread over Northern Europe from Poland to the Baltic sea. They made use of stone and antler as tools. They were familiar with the use of harpoons also.

In pioneer centres between 8,000 and 7,000 B.C. agriculture marked the end of the period of savages and ushered in barbarism. Gradually wild crop gathering was replaced by agriculture and hunting by animal husbandry. But fishing continued to be a vocation till the days of modern civilised man. However aquaculture developed for the controlled of aquatic organisms, especially fish. Experts feel that aquaculture may have had even more remote connections in the highly organised ancient water oriented civilisation of Near East in which fish was an important component of food. Fishing sites conservation of fish, raised in ponds figured in Egyptian bas-reliefs. Pisi-culture was also widely practised in the Roman and Gaul empires. For many centuries the people of Indo-Pacific region and the Chinese have reared fish in plenty.[2] Fish species like Tilpia and Carp had been cultured long before the Christian era by the ancient Egyptian and Chinese people.[3] Aquaculture which was originated as a subsistance level activity thus had its cradle in ancient Egypt and China.

The Chinese saying "Give a man fish and he will eat only once, teach him fish culture, he will eat fish throughout his life", symbolises the philosophy of life that gave to the earliest form of fish farming or husbandry so well described in Chinese 'fish culture classic' written by Fan Li in 460 B.C. The origin of aquaculture of fish farming in China is commonly attributed to Wen Fang, the founder of Chow Dynasty. The last ruler of Shang

---

2. Sarma, B.C., "Social Issues in *Aquaculture*", *Aquaculture Research Needs for 2,000 A.D., Op. Cit.*, P. 399.
3. Allen R.G., Bestford, L.W., Schur, A.M. Johnston, *Bio-Economic of Aquaculture*, Vol. 13, Elsevier Science Publishers, Amesterdam, Netherlands, 1987, p. 1.

Dynasty had Wen Fang confined to an estate in Honan Province between 1135-1122 B.C. There, Wen Feng built one or more ponds, filled them with fish and made the first recorded reference to the behaviour and growth of fish, according to Fan Li. [4] Even the Japanese, modern leaders in aquaculture technology have a long tradition in aquatic husbandry. Mullets were reared in ponds in Hawaian Islands before Captain Cook discovered them. Asia and Far East are therefore said to be the original homes of aquaculture. All of these aquaculture technologies evolved without any benefit of scientific approach of the 20th century. They were only traditional in nature.

Evolution of Aquaculture

From the beginning traditional type of fish farming, aquaculture activity has expanded in area and methods of technology. Although the traditional type of aquaculture is still in practice in more or less the same way in many areas of the world, It is true that aquaculture has undergone considerable expansion, diversification and technological change during the later half of the present century. From the subsistence level activity has become a commercial scale operation with great advantages to the rural economies in particular. Aquaculture has been the main stay of the many villages and community development programmes in Asia and African countries.

Interest in aquaculture has also spread to large private firms and multinational corporations and their entrepreneurship ranges from small scale rural operations to 'vertically integrated corporate undertakings'. The variety of cultured organisms have also expanded, besides the culture of fish, Oysters, Mussels, Clams, Soallops, Lobstors, Seaweeds, Turtles and in high demand Shrimp are began to be reared under captivity.

Types of Aquaculture

Depending on the variations in hydrobiological features, particularly salinity, aquaculture, could be broadly divided into Freshwater aquaculture, Brackishwater aquaculture, Mariculture

---

4. Evan Brown, E., *World Fish Farming : Cultivation and Economics*, AVI Publishing Co., INL Westport, Connecticut, U.S.A., 1977, p. 1.

and Methaline culture. The ranges in salinity in these culture systems are : 0.5 per cent, 0.5-30 per cent, 30-35 and above 35 per cent.

Freshwater Aquaculture

This includes pond fishculture comprising monosex culture, monospecies culture, composite fish culture, air-breathing fish culture, fish culture in tanks, race ways, ornamental fish culture, sport fish culture and integrated fish farming.

The important freshwater culture fisheries include major carps such as catla, rohu, mrigal, common carp, hilsa, mullets, shrimp etc.

Brackishwater Aquaculture

The brackishwater resources along the coastline consist of backwaters, brackishwater creeks, marshy lands, canals and estuaries. Brackishwater area is rich in plankton and oxygen. Fish such as milk fish, chanos, lates, mullets, and shrimp species like P. monodon, P. indicus etc., can be cultured under this system.

Mariculture

This type of culture can be adopted in nearshore, coastal and open sea environment. Floating cages for the culture of fish, shrimp and lobstars, racks for oysters, raff and rope, pole and long lines for mussels, nets or webbings for seaweeds are at present widely employed for this type of culture.

Methaline Culture

This farm of culture is applicable to super saline salt pan areas where salinity is more than 200 per cent. Culture of Brine Shrimp 'ARTHEMIA SALINA' can be undertaken under this culture.[5]

In India the first two types of culture practices are extensively used. Methaline culture is not in operation so far. However, mariculture which is new to India has been extended from Japan in recent years. Mariculture is in practice in India in States like

5. Santhanam, K., *et al.*, *A Manual of Freshwater Aquaculture, Op. Cit.*, pp. 2-6.

Kerala, Tamilnadu and Andhra Pradesh to a negligeable extent. The Central Marine Fisheries Research Institute (CMFRI) has been instrumental in propagating and experimenting mariculture in our country.

Relative Advantages of Aquaculture of Shrimp Over Capture Fisheries

In the recent past, for human beings, science has provided better methods of parental care, better nutrition and prevention or control of diseases. These measures were extended to decrease hum suffering and provide longevity. This resulted in population explosion in the tropical world including India. If the same scientific approach (aquaculture) to fish and crustaceans is applied. a similar population explosion in fish or 'aquaplosion' will happen. Then it will be possible to combat human population explosion with aquaplosion of fish by meeting the nutritional needs of human beings with fish food.[6]

Scientific aquaculture can ensure steady and regular production of fish which may be increased considerably with planned input of modern technology, just like in the case of scientific agriculture.[7]

The shift of population from rural areas to towns and cities, which has created serious socio-political problems can be halted, through properly planned and systematically integrated aquaculture in rural development programmes.

Of all the aquaculture products shrimp is considered to be the most important by virtue of its taste high unit value, persistent demand and universal appeal among the wealthier nations of the world.

Cultured shrimp production has some advantages over capture shrimp fisheries. First of all shrimp catch is seasonal and unreliable even in tropical waters and may fluctuate from year to year. This keeps the shrimp processing industry idle over extended periods and adds to the storage costs of both exporters and importers who must maintain more or less uniform supply of shrimp throughout

---

6. Divedi S.N., "Aquaculture 2000 A.D. Prospects and Perspectives", in K.K. Trivedi (Ed.), *Fisheries Development 2000A.D.*, 1990, pp. 153-154.
7. Bose, A.N., Gosh, S.N., Young, C.T. and Mitra, A., *Coastal Aquaculture Engineering*, Oxford & IBH, New Delhi, 1991, p. 1.

the year.

Cultured shrimp production on the other hand can be planned in advance at least in the tropics and can be run continuously. This improves the efficiency of processors and may reduce the costs of exporter and importer.

Shrimp catch from wild source is determined by geographical and seasonal conditions. But shrimp farmers can select more flexibly from wide range of culturable species and can produce the sizes required, according to the consumers demand.

An additional advantage of cultured shrimp over captured shrimp is that it has nothing to do with Turtles. This is not a joke! The U.S.A. threatens shrimp producing countries with embargo if their trawlers are not mandated to use Turtle Excluding Devise (TED). Shrimp culture does not threaten turtles because their sandy breeding grounds are not suitable for pond construction. So authorities in the U.S.A. may exempt cultured shrimp from 'Turtle embargo'.[8]

Preserving the freshness of shrimp is certainly the most important pre-condition for producing high quality products from cultured shrimp. Cultured shrimp is capable to withstand transport and can be kept alive for more time than the captured shrimp. That is why the processors prefer shrimp to captured one provided there are no flaws in the culturing, harvesting post-harvesting process, because the quality of cultured shrimp depends on the methods used at all these stages.

The major advantage of shrimp aquaculture is that production can be close to the centres of population and the harvesting can be adjusted to the demand. Spoilage losses and distribution problems can therefore be minimised.[9]

Aquaculture of shrimp and other fisheries reduced pressure on the conventional fishery resources, and can be considered as an additional safety net to meet the growing demand for fisheries.

All these advantages, together with the growing urgent food needs of the people have resulted in greater attention being directed

---

8. Imre Casvas, "The impact of aquaculture on shrimp industry", *Infofish International*, No. 1/93, Jan./Feb., Issue No. 127-20212, MC (P) No. 965/5/83, pp. 78/10191, pp. 42-48.

9. David James, "The Future for Fish in Nutrition", *Infofish Marketing Digest*, No. 4/84, ISS. No. 27-2012, MC (p) 155/84, KDN 0864/84, p. 41.

to aquaculture as logical and useful form of fishery production. It is estimated that about 100 to 140 million tonnes of edible fishery products will be required in order to meet the demand of the projected world population by the year 2000 A.D. The maximum expected production is only about 80 million tonnes. Thus there is a deficit of approximately 20 to 60 million tonnes to be made up and probably the only major means presently known for this is an accelerated development of aquaculture.[10] It is expected to contribute around a quarter of the global fishery harvest by the turn of the century.

Relative Position of India in the World Shrimp Culture

Over the last decade, fish production has been growing at a significant pace. While a substantial part of the growth has come from the exploitation of marine fisheries in Latin America, South East Asia and the Pacific region, remarkable developments have radically transformed the aquaculture scene particularly in South East Asia.[11] The geographical distribution of aquaculture development clearly shows that Asia continues to be the major producer, followed by Europe. Although the production from Africa continues to be small in relation to the other regions, there has been some improvement in the last two years. This applies to Latin America as well as North America, indicating that progress is being made, despite the slow start and the deficiencies of infrastructute. From the point of view of commodity profiles, the main increases seems to be in crustaceans and fin-fish for which there is greater consumer demand. There has been a very rapid increase in fish and shrimp production in South East Asian countries like, Philippines, Thailand, China, Indonesia, Ecuador, India etc., under scientific aquaculture. Aquaculture has quickly turned into a flourishing industry in the world through its phenomenal performance. Shrimp culture dominates the world aquaculture scene at present.

The word aquaculture production of shrimp in 1990 was estimated at 6,63,000 tonnes foom a total culture area of one million hectares and 85 per cent of the global farmed shrimp

10. Pillay, T.V.R., Fisheries Development 2000 A.D., *Aquaculture Productivity, Op. Cit.*, p. 154.

11. Srinivasa Sastry, C., "National Policy for Aquaculture Development", *Aquaculture Productivity, Op. Cit.*, p. 21.

comes from Asian countries. China was the largest producer of shrimp in the world followed by Indonesia and Thailand which occupied second and third positions. In 1990 India was in fifth position. The global farmed shrimp production details of 1994 puts Thailand in first position with 2,25,000 tonnes of shrimp production from a total world cultured shrimp production of 7,33,000 M. Tonnes achieved from a total culture area of nearly 1 million, one and half lakh hectares. Ecuadar and China occupies the second and third positions. India with its excellent performance improved its place from fifth in 1990 to fourth in 1994. Thus, India is recognised as one of the world leaders in cultured shrimp production.

Fish culture has been an age old practice in India. There is evidence of such practices in India from the remote past. But the importance of aquaculture as a source of food production and as an industry on scientific lines has been comparatively recent.

The early brackishwater farming undertaken by CIFRI and the CMFRI led to the development of seed and culture technologies of marine species of fish including shrimps and Molluses. A new institute by name Central Institute of Brackishwater Aquaculture (CIBA) was started in 1985, with Madras as its centre. This is a nodal research organisationestablished by ICAR during the 7th Plan period. Central Institute of Fisheries Education (CIFE), Bombay which is now deemed as university has a component of brackishwater aquaculture in all its education and extension training programmes with its farm facility at Kakinada in Andhra Pradesh. Several fisheries colleges were established to promote aquaculture education and to augment the fish production. These colleges are situated at the following places : Mangalore (Karnatak), Tuticorin (Tamil Nadu), Nellore (Andhra Pradesh), etc.

Brackishwater Shrimp Culture in Andhra Pradesh

Andhra Pradesh has a coast line of 982KM. Perennial rivers like Godavari, Krishna and Penna flow through the State before their confluence with the bay of Bengal. Important brackishwater areas of the State are Srikakulam swamps in the north, Godavari estuarine system, Kakinada backwaters, Krishna estuary and Pullicat Lake with major portion in Nellore District in the South. The estuaries, backwaters, bays, and brackishwater lagoons form an important brackishwater culture sources of the State. These eco-systems are highly productive and give rise to highly valuable fisheries such as shrimp, mullet and the migrating hilsa.

Table 3.1 : Global Aquaculture Production of Shrimps-1994

Country	*Head-on Production (Metirc Tones)*	*Hectares in Production*	*Rate of production*
Thailand	2,25,000	80,000	2,813
Ecuador	1,00,000	90,000	1,111
Indonesia	1,00,000	3,00,000	333
India	70,000	90,000	875
Vietnam	50,000	2,25,000	222
China	35,000	1,50,000	233
Bangladesh	35,000	1,10,000	318
Philippines	30,000	50,000	600
Twain	25,000	7,000	3,571
Mexico	12,000	12,000	1,000
Columbia	10,000	2,600	3,846
Honduras	10,000	11,000	909
United States	2,000	7,000	2,857
Brazil	2,000	2,000	2,000
Others	27,000	27,000	1,000
Total	7,33,000	11,47,000	639

Source : World Shrimp Farming, Shrimp News International, San Diego, U.S.A.

First experimental brackishwater farms in the State are located at Kakinada. The farms belong to Andhra Pradesh Argicultural University (APAU) and the Central Institute of Fisheries Education (CIFE). The brackishwater fish farm of APAU at Kakinada was one of the six centres established under Indian Council of Agricultural Research scheme of All India Coordinated Research Project on brackishwater fish and shrimp culture (1976-85). Nearly 67 experiments have been conducted at Kakinada farm on the monoculture of P. monodon, P. indicus, chanos-Chanos, mixed culture of shrimp and fish and polyculture of peneaid shrimp.

A.P.A.U. was the first institute in Andhra Pradesh to initiate detailed experiments on the natural seed prospects. Seasonal availability of the seed of the economically important peneaeid shrimp and other cultivable fish, efficiency of collecting gear, identification of suitable species and their transportation were studied at this centre. Seed calendar was prepared and nursery management experiments were also conducted. And training in these areas were given to the rural youth at this centre.

The Bay of Bengal Programme (BOBP) has also established and organised a small scale fisheries project on experimental shrimp farming at Polekurru near Kakinada. Culture trials with P. Monodon and P. Indicus in different types of trenches, full digging, tidefed, pumpfed etc., were conducted. Thus the recent developments in brackishwater aquaculture of shrimp in particular, had its roots in the results of the experiments of brackishwater farms of A.P.A.U., CIFE and BOBP.

Private Entrepreneurs in Brackishwater Shrimp Culture

The entry of he private enterpreneurs into the brackishwater shrimp culture in Andhra Pradesh has changed the complextion of aquaculture of shrimp under brackishwater conditions. The private entrepreneurs picked up this activity from the point where the APAU, CIFE and BOBP have left. The rush of private entrepreneurs into this venture commenced with the entry of Rank Aqua in 1989. Subsequently, Thapars, Nagarjuna, D.C.L. Maritech, Alsa Marine, Coromondal Aqua, Sharat Sea foods, Sharani-Sindhu shrimp farm, Aquamarine, Water Base, Base Kings International, Suvarna, N.C.C. Bluewater, Bommidalva, Suryachakra, Carewell, Magunta Aqua Farm etc., have entered into this 100 per cent export oriented enterprise.

The entry of these corporate farms created an aquaculture boom in Andhra Pradesh and ushered in the 'Blue Revolution'. The water areas along the coast line which were hitherto considered unproductive became most sought after 'Prime Lands". Shrimp culture is now developing in Andhra Pradesh at a fast pace and large areas of fallow brackishwater lands are being converted intoshrimp farms. Andhra Pradesh stands in the first place followed by West Bengal in the production of cultured shrimp in the country. The phase of transformation in shrimp culture is accelerated in the State through planned infrastructure development, input supplies, and technology transfer. Several foreign technical farms from

Taiwan, Philippines, Japan and Thailand have established links with corporate farms in Andhra Pradesh. The capital invested by the private enterprise in this venture so far is estimated at about 1000 crores. In 1993 about 80 farms have entered into the capital market with public issues in Andhra Pradesh. Among them, the shrimp culture farms toped the list in moblising the public funds.

World Bank Assisted Shrimp And Fish Culture Project

Towards promoting sustainable and increased shrimp production in the State, the project with the financial and technical assistance from the World Bank was taken up by the State Government with an outlay of nearly Rs. 83 crores. About 2000 hectares of brackishwater area would be developed by taking up semi-intensive shrimp culture under this project. 2000 small farmers and 55 commercial farmers are expected to be benefitted through this project. The agreement to this effect was already finalised and signed by the World Bank, the Government of India and the Government of Andhra Pradesh. The project in offing is expected to have greater demonstration effect on the shrimp farming in the State.

Seed Resources

With the increase in the area under brackishwater shrimp culture in the State, the requirement of shimp seed has been increasing. The present demand for seed is estimated at 500 millions, while the present capacity of commercial hatchery of MPEDA *i.e.*, TASSPARC at Visakhapatnam is only 100 million PL-20/annum, production capacity. Therefore, almost the entire requirement of shrimp seed is being collected from natural seed resources. The seed of P. monodon are dominant in northern districts, while P. indicus are available in southern districts of the State. The estuarine and mangrove regions of Godavari and Krishna are major seed grounds, and these natural resources are heavily exploited to meet the seed requirement. Seasonal fluctuations of seed availability, is limiting the culture period and system of culture considering the shrimp seed as a major constraint for the development of shrimp farming. The hatechery at Suryalanka in Guntur District under UNDP assisted project and a number of hatcheries under the Word Bank project are being established in the state to meet the seed requirements. In addition, several hatcheries are coming up under private sector.

Promotional Activities of MPEDA and BFEDA

Marine products Export Development Authority (MPEDA), a nodal agency in the promotion of aquaculture of shrimp in India has opened three regional offices at Visakhapatnam, Vijayawada and Bhimavaram in the State to encourage and develop brackishwater shrimp culture. Besides providing training and technical guidance to the farmers, MPEDA also implements limited subsidy schemes. The hatchery established by MPEDA with 100 million PL-20 capacity is the single largest government hatchery seed source in Andhra Pradesh.

Considering the fact that there is an immense scope for the development of brackishwater shrimp farming, 37 Brackishwater Fish Farmers Development Agencies (BFFDA) have been sanctioned by the Centre of which six are set up in Andhra Pradesh at Nellore, Machilipatnam, Kakinada, Ongole, Bhimavaram and Srikakulam. BFFDA provide package of inputs, technology and financial support to the farmers selected under its scheme.

Organisational Support

The government of Andhra Pradesh has recognised the brackishwater shrimp aquaculture as a high priority area for increasing shrimp production. The government has decided to lease out 17,000 hectares brackishwater government land suitable of for aquaculture at the ratio of 60 per cent to the fishermen co-operatives, 20 per cent to self-employed technocrats and 20 per cent to progressive enterpreneurs. In the first instance 3000 hectares (100 hectares each in Nellore, Krishna and East Godavari districts) was notified for leasing out in 1989. An extent of 8000 hectares was notified in 1991 to be leased out in the second phase. Beneficiaries for allotment in the first phase has already been selected. The second phase is in progress. The government is also directly implementing shrimp culture schemes at different centres through DRDA, BFFDA and SC-BC Corporations for the benefit of weaker sections of he society. The non-governmental organisations like CASA, FACT etc. are also helping the small and marginal farmers engaged in shrimp culture in securing input technical guidance and bank loans.

Banks in Brackishwater Shrimp Culture

Nationalised banks and cooperative banks with the participation of National Bank for Agricultural and Rural Development (NBARD) are extending adequate financial support to this sector. Insurance Companies are also planning to extend crop insurance schemes to the shrimp farmers in the State.

Table 3.2 : State-wise Details of Shrimp Farming in India.

State	*Estimated Potential Area (ha)*		*Year*			
			1990-91	*1991-92*	*1992-93*	*1993-94*
West Bengal	4,05,000	A	33,815	33,918	34,050	34,150
		P	12,500	13,800	16,300	16,500
Orissa	31,600	A	7,075	7,417	7,760	8,150
		P	4,100	3,800	4,300	3,300
Andhra Pradesh	1,50,000	A	6,000	8,100	9,500	19,500
		P	7,350	9,700	12,800	26,000
Tamil Nadu	56,000	A	250	480	530	1,050
		P	450	700	1,100	2,000
Pondicherry	800	A	Neg.	Neg.	Neg.	Neg.
		P	Neg.	Neg.	Neg.	Neg.
Kerala	65,000	A	13,000	13,145	13,400	13,860
		P	8,925	9,500	9,750	11,500
Karnataka	8,000	A	2,500	2,52	2,570	2,600
		P	1,000	1,100	1,150	1,500
Goa	18,500	A	525	525	550	575
		P	245	300	350	400
Maharashtra	80,000	A	1,800	1,869	1,980	2,180
		P	800	930	1,050	300
Gujrat	3,76,000	A	125	231	360	475
		P	125	170	200	500
	11,90,000	A	65,100	68,227	70,700	87,540
		P	35,500	40,000	47,00	62,000

A- Area under culture. P. Estimated Production.

Source : 'Marine Products Exports Development Authority'. Kochi, Kerala.

Enterprising and Adoptable Nature of Andhra Pradesh Farmers

The farmers of Andhra Pradesh are widely known for their acumen, adaptability and outstanding achievements in the field of agriculture. Their contibution in the phenomenal success of 'green' and white revolutions acknowledged and appreciated nation-wide. The same class of farmers are now all set to bring about the 'Shrimp Revolution' in the State.

According to E.G. Silas, Andhra Pradesh would be the major producer of cultured shrimp in India by the end of this decade. In the opinion of G. Sanathan Kumar, very soon Andhra Pradesh will be the second Tawian in shrimp culture. With vast brackishwater potential, technical viability, favourable environment and enterprising farmers, Andhra Pradesh is expected to be a centre of 'Shrimp Revolution' in the country very soon.

Chapter 4

Review of Literature

Having examined in brief the genesis of aquaculture, its importance and present status both at national and state level with special reference to brackishwater shrimp culture, it is appropriate to critically evaluate the empirical studies conducted in economics of aquaculture in India in order to stipulate the specific problem or area under the present study. This chapter presents the review of select literature in this area of research.

Agricultural allied activities in general and aquaculture of fisheries (fish, shellfish) in particular have attracted little attention from the planners and the researchers in India in the past. Specially, the brackishwater aquaculture is almost left uncovered or untouched. It is only in the recent past that the researchers and planners got awareness that aquaculture of fisheries can be a potential source of protein rich food, employment and foreign exchange and that it can also contribute in mitigating the problems of foreign exchange and unemployment.

Though significant research work has been done in biological and other marine sciences it is not properly integrated or coordinated with economics. Research conducted on fisheries so far was related to the problems such as unorganised market structure, high feed prices, low fish prices, fish diseases, mechanisation and fish landings and inland fish culture etc. Many of these studies have failed to make any specific attempt to examine in detail the economic aspects of brackishwater shrimp culture.

However, a brief review of these studies is attempted in this chapter. The review that follows is selective and purposive. The literature in this context is reviewed under the following three heads :

(1) Growth and importance of aquaculture of fisheries;

(2) Marketing aspects of fisheries and

(3) Cost and returns in culture fisheries.

Growth and Importance of Aquaculture of Fisheries

Srivastava *et al*[1] in their study have attempted to find out whether the actual field conditions match with research results or not and to suggest future strategy to develop brackishwater fish culture. In this study an attempt has also been made to compare the economics of intensive and traditional type of brackishwater culture, to bring about regional differences in management and to assess the risks involved in brackishwater aquaculture.

The survey under this study encompasses complete brackishwater fishery system. Using available publications and secondary data, places were selected for survey. The survey covered the fish farmers across the coastal states of Kerala, West Bengal, Karnataka, Andhra Pradesh, Tamil Nadu and Orissa. The data was collected from 9 intensive, 35 extensive and 45 traditional (filtration method) farmers. The experiments conducted in all the three types of brackishwater farms show that brackishwater farms can attain 82 per cent return on investment.

According to this study, intensive culture is a recent scientific innovation. Therefore only a few farmers have adopted it and only 20 per cent of them continued it while the remaining reverted to the traditional method. Shrimp production was the highest in all the three brackishwater culture methods. Gross revenue per hectare is highest in the intensive culture and lowest in filtaration fields due to the differences in the variety and size of shrimp. This study points out that the technical and managerial problems, and the conflicts in ownership of brackishwater areas are the important regional constraints for the development of brackishwater culture.

The study suggests that the supply of water through pipe system, implementation of long term leasing policy, adequate

1. Srivastava, U.K., *et al.*, "Economics of Brackishwater Culture in India : Some Preliminary Evidence",in U.K. Srivastava and M. Dharma Reddy (Eds.), *Fisheries Development in India*, Concept Publishing Company, New Delhi, 1983, pp. 391-409.

supply of seed, finances and engineering assistance would enable the farmers to get more yield. The study further suggests that a strong extensive service system aimed at making available the results of new technology to the actual users in the field is urgently needed. But the data used in this study are quite inadequate for any policy measure to be taken up.

Sarma has made an attempt to impress that social issues in aquaculture are as important as other issues.[2] This point has been emphasised by illustrating the socio-economic study of Sunderbans in this study.

Sundarbans are at the confluence of Ganga-Brahmaputra river system. It is the largest river delta system in the world comprising about 54 islands. This study is confined to lower Sunderban delta where there are 108 households including fishery and non-fishery farmers. In this study, the farmers were interviewed using various parameters like family income, expenditure pattern, housing conditions, land options etc.

The study was conducted on specific hypothesis that income from fisheries activity is greater than that from agricultural operations. The study reveals that the gross rate of return on fishery was 35 per cent higher than that of agriculture. The study also points out that calorie intake of fisheries households is much better than non-fishery households. It is interesting to note under this study that out of 108 households 56 had their first preference for agriculture and 52 for aquaculture, even though the gross return for fisheries was much higher.

The study concludes that farmers owning agricultural lands have better status in the society than fishermen and fish farmers. Therefore, the majority of farmers prefer agriculture to aquaculture. The study suggests that involvement of women in aquaculture should be encouraged and social issues in fisheries should be highlighted to the extent they deserve.

Subba Rao,[3] made an attempt to make a multidimentional study of fisheries with particular reference to Andhra Pradesh. He

2. Sarma, B.C., "Social Issues in Aquaculture", in K.K. Trivedi (Ed.), *Aquaculture Research Needs for 2000 A.D.*, Oxford & IBH Publishing Company, New Delhi, 1988, pp. 399-405.

3. Subba Rao, N., *Economics of Fisheries : A Case Study of Andhra Pradesh*, Daya Publishing House, New Delhi, 1986.

outlined the history of fisheries and various sources of fisheries in Andhra Pradesh. Besides pin-poining the various trends in fish production (production of different species of fish) under various five year plans, he dwelt upon the fishing technology, its advantages and drawbacks.

According to this study inadequate financial resources allotted to fishery sector, failure of implementation of government schemes in the stipulated time, conservative nature of fishermen in adopting new methods and the lack of extention service facilities are the reasons for the failure of fisheries planning in Andhra Pradesh. In the back drop of the findings of this study, the author has suggested the formation of a study group to go into various aspects of fisheries development and to make necessary suggestions for the removal of the current obstacles.

Though this study highlights various economic aspects of fisheries like trends in production cost and returns, demand and supply projections, exports technology etc., it fails to throw light on the economics of brackishwater culture fisheries. It only makes a passing reference to the brackishwater culture fishery potential in the State of Andhra Pradesh without going deep into various related aspects.

Sandip Mathur and Gopala Krishna have conducted three studies on shrimp culture holdings, two studies in East Godavari and one in West Godavari District of Andhra Pradesh.[4] The first case study is related to Damisetty Prawn Farm located in Chinnavakala Polekurm village of East Godavari District. This study observes that the shrimp culture is commercially viable and the local value of land has gone up by 10 times within a span of six years, because shrimp culture has been successful in this village.

The second study is related to District Rural Development Agency (DRDA) farm at Kulkharmband in East Godavari District which is a unique case of co-operative management of small scale private holdings. The profits earned under this type of farming constitute returns to personal and to individual management and labour.

---

4. Sandip, K. Mathur and Gopala Krishna, G., "Economics of Commercial Brackishwater Farming : Case Studies in Andhra Pradesh", *Aquaculture Productivity*, Oxford and IBH Publishing Pvt. Ltd., New Delhi, pp. 267-275.

The third study refers to Sagar Prawn Farming located in Pedaguru village, West Godavari District of Andhra Pradesh. In this farm, shrimp culture was undertaken in the first crop and Indian major carps in the second crop with the change in soil salinity. The study suggests that enterpreneurs at present need constant coordination and guidance in culture practices. All the three case studies emphasise the success of private entrepreneural activity in large scale as well as in the government aided small scale aquaculture enterprises.

Radhasyam and Tripathy,[5] analysed the experience of a village, Nakhurpatnam, in Bhuvaneswar block of Puri District of Orissa State under composite fish culture. In 1978, when the villagers were not aware of aquaculture of fisheries, 88.64 per cent of the families in this village lived below the poverty line with a monthly income of Rs. 185 per family. After 1978, they adopted culture method of rearing fish and started getting good returns. In 1990 *i.e.*, after 12 years when the survey was conducted, the average monthly income of each family increased to Rs. 12,863.43. The net return from fish culture was the highest *i.e.* 89-19 per cent while it was between 1.13 to 4.31 per cent from other activities. The per centage of families below the poverty line came down to 45.45 per cent.

The experience of this village shows that acquaculture diversified in many ways is capable of contributing significantly towards a qualitative shift in the socio-economic status of the rural poor by enlarging the scope of self employment.

On the basis of this experience the researchers suggest that rural human resources can be gainfully employed in aquaculture. It is a socio-economic study which portrays the situation of Nakhurpatnam village over a period of time. It ignores the implications of aquaculture of fisheries under the changing bio-technical climate at a particular point of time.

Evan Brown[6] attempts to provide economic and technical information by countries and by species-wise on commercial

---

5. Radhasyam, Tripathy, N.K., "Aquaculture as a nucleus for integrated rural development—An experience", *Fishing Chimes*, Dec. 1992, pp. 37-41.
6. Evan Brown, E., *World Fish Farming : Cultivation and Economics*, University of Georgia, The AVI Publishing Co., INC, Westport, Connecticut, U.S.A., 1977.

aquaculture and tries to fill the existing gap in the literature on commercial aquaculture. The main objective of this study is to present aquaculture arts and to answer some important questions on aquaculture in 28 countries in easy to read fashion.

This study makes a very sincere attempt to present economic data on most recent cost relationships, marketing margin, producer and consumer prices and data on imports and exports wherever possible.

Aquaculture data including feeding practices, foods and types of production facilities for freshwater, brackishwater and marine production have been provided in this study. Data are presented in both matric and American systems to allow maximum understanding. The study also covers a wide range of subjects related to private and public aquaculturists.

Any data will become useless after a period of time. So are the data provided in this study. However, the data presented in this study serves as a benchmark for future comparisons. The study emphasises that the increase in fish production is significantly related not only to marketing and price structures but also to political and legal considerations.

According to Rao's [7] study the growth of fisheries in India is on a lower scale when compared to advanced fishing nations such as Peru, Japan and USSR. The study finds that the average fish production in India during 1950-65 was only 3.8 per cent. The low rate of growth of fish production is attributable to the primitive methods of fishing followed in our country. In the opinion of Rao mechanisation programme of fishing has not picked up momentum as the fishing industry in the country is not properly orgnised.

Rao also discusses the increasing demand for different species of fish. He points out that the demand for fish is increasing abnormally owing to its importance of food value for cattle and poultry. But the per capita consumption of fish in India is very low when compared to advanced countries. He concludes that development of fisheries is going to play a crucial role in India's national food economy and feels that to increase fish production, modern methods of fishing have to be introduced.

---

7. Rao, P.S., "Some aspects of Economics of Fisheries in India", *The Indian Journal of Agricultural Economics*, Vol. 23 (4), p. 226.

Marketing Aspects of Fisheries

The studies which have addressed themselves to the issues of marketing of fisheries can be broadly grouped as follows :

Profitability of fish culture mostly depends upon cost of production, market prices and marketing costs under a number of intermediaries operating in the market. The efficiency of marketing system depends upon the net margin of profit retained by the farmers. Some studies which dealt with the important aspects of fisheries marketing are reviewed hereunder.

Srivastava's[7a] study aimed at examining the present status of aquaculture in India, economics of both freshwater and brackishwater aquaculture, marketing of produce and research needs in the area of economics and marketing of aquaculture products.

According to this study marketing is the most essential and important link in the success of aquaculture, but it is the most neglected activity. Remunerative price for the producer and reasonable price for the consumer can be assured only by strengthening the fish marketing structure. Price realisation to the fish farmer in domestic fresh fish is very low (maximum of Rs. 9 per Kg.) mainly because most of the fish ponds are located far from the consumer centres and the farmers are exploited by the traders. Fish farmers are under pressure to sell their produce at pond site at whatever price they realise because of lack of proper marketing infrastructure. Very little effort has been made by the fisheries cooperatives and corporations in this area. Research efforts are required to suggest means of regulations of market intermediaries and management of terminal markets according to this study.

Subba Rao and Chowdary,[8] in their study presented marketing costs and profit margins received by the producers at different local markets like Guntur, Vijayawada, Eluru and Kakinada in Andhra Pradesh and Howrah market in West Bengal. For the

---

7a. Srivastava, U.K., "Aquaculture : Marketing and Economics in India", *Aquaculture Research Needs for 2000 A.D.*, Oxford & IBH Publishing Co., New Delhi, pp. 311-325.

8. Subba Rao, D.V., and Chowdary, K.R., "A Study of Marketing Factors influencing prices of Inland Fish in selected Markets", *Indian Journal of Agricultural Marketing*, Vol. 2 (2), 1988, pp. 170-175.

purpose of this study 59 fish producers and 24 middle men were selected from these markets by adopting stratified random sampling technique. The prices prevailing at the end of the harvest season, *i.e.*, March-April, 1982 alone were examined.

The study discusses, marketing costs incurred on packing material, labour, transport, commission, discount etc., by the middle men both in the local markets and Howrah market and per centage share of different marketing costs from consumer's rupee for different varieties, viz., Catla, Rohu and Mrigal. The total marketing costs borne by the middlemen per Kg. of fish amounted to Rs. 0.81 in the local markets and Rs. 7.67 in Howrah market irrespective of the species. But the percentage share of marketing costs with consumer's rupee varied for different species due to variation in consumers price. The per centage share of marketing costs in the consumers rupee are 8.74, 9.85 and 12.78 for catla, Rohu and Mrigal respectively in the local markets, while it is 48.12 and 45.87 per cent in Howrah market for Catla and Rohu respectively.

This study has also attempted to discuss the marketing costs to find out the share of various functionaries in the consumer's rupee.

According to this study the net share of producer in consumer's rupee accounts for 70.3, 69.7 and 67.0 per cent for Catla, Rohu Mrigal respectively in the local markets, while it is 40.9 and 34.3 per cent for Catla and Rohu respectively in Howrah market. The profit margin for middle men is uniform *i.e.*, about 20 per cent in local and Howrah markets for all varieties except for Catla which was 11 per cent in Howrah market.

The study suggests the adjustment of harvesting during the months of May and June to realise better prices to producers and to harvest the fish when its body weight is 1 Kg. and above. The study finally suggests the sale of fish produce through co-operatives in order to avoid middle men and to get increasing margin of profit.

Raju et al[9] in their study examine the sale procedure and price spread in fish marketing to assess the marketing efficiency.

---

9. Raju, D.S.N., Krishnaiah, J., Subba Rama Raju, K., and Jagannadha Raju, "Marketing of Inland Fish from West Godavari, Andhra Pradesh : Problems and Policy Issues", *Indian Journal of Agricultural Marketing*, Vol. 3 (1), 1989, pp. 74-77.

According to them the major share of the consumer rupee goes to the producer (60 per cent), followed by the retailer (16.24 per cent), and the wholesaler (8 per cent). In marketing cost, transport charges, market fee and commission charges worked out at 3.76 per cent, 3.00 per cent and 3.00 per cent, respectively. The total expenses incurred by the producer accounts for 13.08 per cent of consumer price. The price realised by the producer is Rs. 18,000 out of Rs. 30,000 paid by the consumer for 1 tonne of fish. The study concludes by saying that the existing marketing channel is reasonably efficient because of the higher share of the producer in consumers rupee.

Pradeep, Katiya and Ravish Chandra[10] have evaluated the operational efficiency as a measure of fish marketing efficiency. Their study was conducted with the following objectives : (i) to compare price spread for different sized corps and catfishes to measure operational efficiency, (ii) to calculate the correlation coefficients between market prices for different sized carps and cat-fishes to assess pricing efficiency and (iii) to compare operational and pricing efficiency as a measure of marketing efficiency. The data collected relate to August, 1986 to June, 1987. Two market channels that existed in the study area are identified as (i) Fishermen-Wholesaler-cum-Commission Agent-Retailer-Consumer, (ii) Fishermen-Retailer-Consumer. The study reveals that major share of catch passes through the first channel.

The study shows that fish marketing system in Allahabad is fairly integrated, so far as intermarket price movements are concerned. But, higher retailers profit margin accounts for quite a large proportion of price paid by the consumer. This is a clear symptom of inefficiency.

Finally, the authors suggest that fishermen should get at least minimum remunerative price. This would help in increasing fish production. The monopoly of buyer at wholesale market and supply side at retail market should be checked to establish fair price for the fishermen and consumer. This could be done by encouraging the marketing of fish through co-operatives and State Trading Corporation. The fishermen should be provided with uptodate market information. The study further suggests that the

---

10. Paradeep, K., Katiya and Ravish Chandra, "Fish Marketing Efficiency : A Case Study of Allahabad Fish Market", *Fishing Chimes*, Oct., 1990, pp. 21-29.

fishermen should be supplied with adequate and timely institutional credit facilities to enable them to be free from the clutches of fish traders.

Ramana et al[11] in their study have attempted to identify various marketing channels of fish to work out marketing costs and margins for different varieties of fish and to study the economics of conversion of fresh fish into dry fish.

For the purpose of this study they have selected Bapatla Mandal in Guntur District of Andhra Pradesh which is situated along the coast-line. Forty fishermen, three wholesalers and twelve retailers were selected from three villages in collecting data by using proportionate random sampling technique. The study identified three different fish marketing channels in the study area. They were : (i) Producer-wholesaler-retailer-consumer;(ii) Producer-retailer-consumer; (3) Producer-consumer. The fish sold in the first channel was 82.5 per cent while it was only 12 per cent and 5.2 per cent respectively in the second and third channels.

The marketing costs and margins worked out for different varieties of fish show that margin accrued to the retailer is the highest followed by wholesaler. The producer received a better price and share in the margin of profit in channel two.

The study reveals that 82 per cent of fish producers are illiterate and ignorant of the prevailing market prices. Majority of the producers invariably depend upon wholesalers for financial assistance owing to their poor economic viability which results in exploitation of producers in the product market.

The study finally suggests that the fishermen have to be provided with cold storage, institutional credit facilities, construction of drying platforms and minimum support price to the fish produced by them.

Hari Prakash,[12] made an attempt to study the economics of inland fish culture in West Godavari District of Andhra Pradesh. Out of seven mandals where there was fish culture, three mandals were chosen for this study. In these three mandals there were 360

11. Ramana, U.K., Raghuraman, P., Seetharaman, S., and Eswara Prasad, Y.,"Fish Marketing in Coastal Mandal-A Micro Analysis", *Indian Journal of Agricultural Marketing*, Vol. (2), 1990, pp.189-208.

12. Hari Prakash, P., "Economics of Inland fish Culture : A Case Study of West Godavari District of Andhra Pradesh", Unpublished Doctoral Thesis, Submitted to Andhra University, Visakhapatnam, 1991.

fish farmers. Out of this 50 per cent of the farmers *i.e.,* 180 were selected by using proportionate random sampling technique and the questionnaire was canvassed among them.

According to this study the common problem of all categories of fish farmers is shortage of feed and highfeed prices. This is mainly responsible for the high cost of farming in the study area. It is also observed in this study that farm size and investment are directly proportional. In the total cost of production variable cost constitutes 74 per cent. Among different cost items feed cost is the highest with 32 per cent of the total cost. Production function analysis of this study reveals that in fish farming feed, organic and inorganic fertilisers are the important factors that affect gross returns. The study further reveals that fish farming in the study area is a male dominated occupation as male labour accounts for about 88 per cent of man days. There is no serious marketing problems in the study area. However, the study suggests that marketing, credit and storage facilities have to be further improved to achieve full benefits of fish farming : Though this study is somewhat comprehensive on economics of fisheries, it has not properly explained the concept of aquaculture or fish farming and various types of fish culture.

Costs and Returns in Culture Fisheries

Profitability of aquaculture of fish as in any other enterprise mainly depends upon the costs involved and income obtained by the fish farmers. Therefore, the aspect of costs and returns is considered as an important component of aquaculture activity. An attempt is made in this context to review a few studies which deal with the costs and returns in fish culture.

Dhondyal and Singh[13] in their study have estimated the share of fixed capital in the distribution of overhead cost on the basis of opportunity cost of the cultivated land diverted to culture pond and was roughly calculated at Rs. 15 per acre.

The authors have given an account of the cost of cultivation of fish under systematic cultivation for two years and under natural fishery for one year crop. The total cost of cultivation per annum is Rs. 761.53 per acre in systematic cultivation and Rs. 52.00 per annum under natural fishing. The net income under systematic

13. Dhondyal, S.P., and Singh, G.N., "Benefit cost Ratios in Fish Culture", *Indian Journal of Agricultural Economics*, Vol. 23 (4), 1968, pp. 239-242.

fish culture is Rs. 2,305.13 and under natural fishing it is Rs. 349.56. But the benefit cost ratio is nearly double in the natural fishery because of low level of investment cost. However, from an overall culture point of view. systematic fish culture, owing to its high income generating capacity per acre affects that the high benefit cost ratio and the low cost of production per quintal obtained under natural fishery.

The Commissionerate of Fisheries, Government of Andhra Pradesh[14] estimated the cost and returns of fish culture, and the cost of construction of fish pond for one acre of water spread area. The cost of construction of pond, including digging and trenching worked out at Rs. 16,000. The cost for inlet and outlet construction is estimated at Rs. 10,000. The variable cost for one acre of water spread area is assessed at Rs. 10,000.

According to this study the production of one acre of pond is estimated at 1.6 tonnes and the cost of production per Kg. Rs. 15, thus making a total value of Rs. 24,000. The net returns calculated on the basis of variable cost therefore is Rs. 13,000 per acre of pond.

Padmavathi and Anjaneyulu[15] have made an attempt to study the costs and earnings of three carp culture ponds with different managerial practices in Krishna District of Andhra Pradesh.

For the purpose of this study they have selected three villages namely Bhujabalapatnam (Pond 1) Bhairavapatnam (Pond 2) and Potumarru (Pond 3) in Krishna District. The extent of the Ponds 1, 2 and 3 was 4.0, 6.4 and 1.6 hectares respectively.

The data on the operational costs under major items such as cost of seed, feed, fertilisers etc., for the two successive crops of the three ponds, were computed for one hectare per one year. The data on fish yields and incomes were also calculated for two successive crops.

The data on fish yields, incomes, operational costs, net incomes and returns for the two successive crops showed that in the first crop, the high per centage of return was obtained in Pond 2. It was

---

14. Government of Andhra Pradesh, Commissionerate of Fisheries, *Information Pamphlet*, Hyderabad, 1987, p. 6.

15. Padmavathi, P., and Anjaneyulu, P.S.R., "A Study in the costs and earnings of three carp culture ponds with different managerial practices in Krishna District of Andhra Pradesh", *Aquaculture Economics*, Asian Fisheries Society, Mangalore, pp. 39-41.

due to higher inputs in the form of feed, fertiliser and seed. In Pond 2 the percentage of return was less by about 7 per cent when compared to that of Pond 3 which was relatively good and was contributed by higher inputs in feed, seed and fertilisers.

In the second crop, in Pond 3, a maximum of155.86 per cent of return was achieved. Even though the net income increased by Rs. 75,921 from that of the first crop, it did not reflect in the percentage of returns. In Pond 2 in the second crop, there was substantial drop in return and net income in spite of higher inputs in feed and fertilisers. This was mainly due to low stocking density and the presence of a large number of unwanted fish in the pond.

In Pond 1, in the second crop, the percentage of return (44.84) came down sharply from the previous crop. The production of carps was not commensurate with the increase in inputs, and the net income increased only marginally. This low yield in this pond was due to low oxygen conditions prevailing during most of the culture period.

The study concludes that high net income can be achieved with high inputs in feed, fertilisers and seed, and by adopting intermediate harvesting technique. Measures to eliminate unwanted fish is very essential to improve the yields. According to this study if proper environmental conditions are maintained and steps to eliminate unwanted fish and diseases are taken along with proper use of inputs in feed, fertilisers and seed, the yield, the net income and the percentage of return would greatly improve in carp culture.

Blanco[16] in his study portrays the economic trends of coastal aquaculture in the Philippines. The economic aspects of finfish, skullfish and seaweed farming in the coastal zones of the Philippines including the levels of capital investment, cost of operation of different types of coastal aquaculture, cost-benefit ratios of each method compared to agriculture and the present state of aquaculture development are discussed. The data collected for this study indicate that sometimes the benefits derived through fish farming are more than the benefits received for agriculture. The investment return ratio is also comparatively higher in coastal aquaculture practices in the Philippines according to this study.

---

16. Blanco, G.J., "Economic Trends of Coastal Aquaculture in Philippines", *Coastal Aquaculture in the Indo-Pacific Region*, Fishing News Books, London, 1972.

Upadhyaya and Dutta[17] made an attempt to study the economics of pisciculture operations in Assam. The study was conducted in two blocks namely Jugijan and Gabharu in the districts of Nagaon and Sonitapur. Two villages from each of the blocks have been selected on the basis of accessibility and availability of the number of progressive fish farmers. Lists of progressive fish farmers have been drawn up and from each village 5 farmers have been selected for the study following the random sampling method. Thus in all 20 fish farmers are selected from villages spreading over the two blocks in the two districts.

On the basis of stocking density and on the type of fish culture practised the ponds were classified into two groups *i.e.*, group 'A' ponds and group 'B' ponds. Under group 'A' there were 14 ponds in which Indian major carps viz., Rohu, Catla and Mrigal were cultured. In the remaining 6 ponds under group 'B', composite culture with the Indian major carps and common carp were taken up. The average stocking density was more in 'A' group ponds i.e., 15,160 fingerlings per hectare and 12,532 per hectare in the case of 'B' group ponds. The average investment per hectare of water area has been Rs. 3,823 in the case of group 'A' ponds and Rs. 3,974 in the case of group 'B' ponds. The average fish production recorded per hectare of water area was 808 Kgs. and 1004 Kgs. respectively from group 'A' and 'B' ponds and the net income obtained per hectare from group 'A' and group 'B' ponds was Rs. 17,006 and 22,285 respectively. The cost of production per Kg. of fish was Rs. 4.73 and Rs. 3.95 against the average market rate of Rs. 25.78 and Rs. 26.15 in the two categories of ponds.

The net return on investment from group 'A' and 'B' ponds was 444 per cent and 561 per cent respectively over a period of 12 months. The study reveals that the modern techniques of scientific fish culture are yet to make a significant impact among the rural fish culturists in Assam.

The authors are of the view, that due to ever increasing demand and very high market price for fish, fish culture offers a lucrative source of income to the people of Assam. Though the use of improved techniques, fish culture is likely to play a phenomenal role in the economic development of Assam in the near future.

---

17. Upadhyaya, S.N., and Dutta, A., "Economics of Piscicultural Operations in the Rural Areas of Assam", *Fishing Chimes*, May 1993, pp. 47-50.

Summary

From the review of the above studies the following important findings are observed :

(1) Most of the farmers prefer aquaculture of fish to agriculture, since aquaculture of fish, particularly shrimp, is found to be more income yielding and profit generating than agriculture.

(2) Modern techniques of scientific aquaculture are yet to make a significant impact on the rural aquaculturists who are engaged in extensive type of culture operations. Technical and managerial problems are identified as the main constraints of development of aquaculture activity in the country.

(3) Marketing is the important link in the success of aquaculture enterprise, but it is identified as the most neglected area. The farmers are subjected to exploitation by the wholesalers, retailers and commission agents in almost all channels of marketing and the farmers are unable to get minimum remunerative price for their produce. However, no serious problems are identified in the marketing of shrimp.

(4) Profitability in aquaculture of fish like in any other enterprise depends upon the cost incurred and the income obtained. Therefore the study of costs and returns seem to constitute an important aspect of aquaculture growth.

(5) The studies further reveal that net income under systematic and scientific aquaculture is much higher when compared to the natural or capture fishing, and investment in aquaculture directly relates with high returns.

The literature reviewed on economics of aquaculture of fisheries reveals the paucity of empirical studies and weak research base of economics of brackishwater shrimp aquaculture. The scholar therefore felt that the issues such as costs and returns economics, input-output relationship, market structure, factors inducing the farmers to take up shrimp culture and the socio-economic profile of culture farmers, employment generation in shrimp culture and other issues related to brackishwater shrimp aquaculture need indepth region specific studies. Hence an attempt is made in the study to fill some of the above gaps in this area of research.

Chapter 5

Socio-Economic Characteristics of Sample Brackishwater Shrimp Culture Farmers

Introduction

For a proper study of the structure of brackishwater shrimp farming, it is relevant to understand the socio-economic characteristics of brackishwater shrimp culture farmers, such as their literacy level, caste composition, occupational distribution and family particulars. Before doing so, a brief profile of Krishna District has to be presented to know of the backdrop.[1]

Profile of Krishna District

Krishna District is one of the coastal districts of Andhra Pradesh with its headquarters at Machilipatnam. This district was formerly called as Machilipatnam District. Later it was renamed as Krishna District after the holy river Krishna in 1859. Krishna District was further divided into Krishna and West Godavari Districts in 1925.

The district is located between 15°-43′ and 17°-10′ of the Northern Latitude and 80°-33′ of Eastern Longitude which extends over an area of 8727 square Kms. With a coast line of 88 Kms.

The district is surrounded on the East by Bay of Bengal and West Godavari District, on South by Bay of Bengal, on West by Guntur and Nalgonda Districts and on North by Khammam District.

1. The profile of the district and the selected villages were mainly compiled from the census report of 1991 and Hand Book of Statistics Published by the Chief Planning Officer, Machilipatnam.

Physical Features

The districts is divided into 50 Mandals covering 4 Revenue Divisions viz., (1) Bandar, (2) Vijayawada, (3) Gudivada, (4) Nuzividu. There is one Municipal Corporation Viz., Vijayawada and five Municipalities viz., (1) Machilipatnam, (2) Pedana, (3) Gudivada, (4) Nuzividu, (5) Jaggayyapeta. There are 1003 villages with 923 Panchayats in the district.

The districts is divided into two natural zones viz., (1) Delta and (2) Upland zones. All the Mandals existing in Bandar Division, Gudivada Division and a part of Vijaywada Division (Penamaluru, Kankipadu, Thotlavalluru and a part of Vijayawada rural) comprise the delta and all the Mandals existing in Nuzividu Division and all the remaining Mandals of Vijayawada Division constitute the uplands. The uplands consist of undulating plains broken by low range of eastern ghats. The main source of irrigation in the uplands is tanks. It is being benefitted by the water of Nagarjunsagar also. The delta lands are being irrigated by canals of river Krishna.

Demographic Characteristics

The population of the district is 36.99 lakhs as per the 1991 census, constituting 5.6 per cent of the State while the geographical area of the district is 8727 sq. kms., which is only 2.9 per cent of the area of the State. The district has a population density of 423 per sq. km. as against the State's density of 241 per sq. km. The total population of the District is 36,98,833, out of which 18,78,784 are male and 18,20,044 are female. Of the total population Scheduled Castes constitute 16.56 per cent while the Scheduled Tribes constitute 2.45. The district has a working force of 13,47,044 constituting about 36.40 per cent of the total population. Among the labour, agriculturists constitute 6.56 per cent while agricultural labour constitute 20.27 and the balance of 9.57 are engaged in housing and in other industrial activities.

Climate

Tropical climatic conditions with extreme hot summer and cold winter prevail in this district. April to June are the hotest months with high temperature in May. The monsoon breaks in the middle of June and brings good rains upto the middle of October.

The normal rainfall of this district is 1028.1 mms., 2/3rds of which is received through South-West monsoon.

Soils

There are three types of soils in the district viz., (1) Black Cotton Soil which constitutes 57.6 per cent. (2) Sand-Clay Loams with 26.3,and (3) Red Loams with 19.4 per cent are prevalent in this district, while a small belt of sandy soils constitute 0.7 per cent fringes on the sea coast.

Land Use

The district has 21, 73, 782 acres of land endowed with rich variety of soils. The district occupies an important place in agriculture. The cropped area in the district is nearly 17.5 lakh acres. Paddy is the main crop followed by black gram, greengram and groundnut. Of late the district is forging ahead in shrimp culture in the brackishwater areas of the coastal mandals Kaikaluru, Kalidindi, Bantumilli, Kruthivennu, Konduru and Nagayalanka. Nearly 8000 hectares of land is likely to be brought under brackishwater shrimp culture very soon. Krishna delta system and Nagar junasagar project are the main sources of irrigation.

Rivers

The famous and sacred river Krishna flows in this district. Like Godavari and Cauvery it flows almost across the peninsula from west to east and finally joins the sea at two places. Among the other streams and tributaries existing, Muniyeru, Budameru and Tammileru are significant. Part of the upland mandals as well as the delta mandals are now having the facility of utilising river Krishna's water for irrigation purpose.

Muniyeru is the chief tributory of the river Krishna and flows from north to south across the erstwhile Jaggaiahpeta and Nandigama taluks. This is an additional irrigation source available in this district. Budameru is another hillstream which flows through several Mandals viz., Mylavaram, Vijayawada, Gannavaram, Gudivada, Mandavalli, Kaikaluru and finally falls into Kolleru lake. Agriculturally the importance of he above sources is quite high as they not only feed minor irrigation source and ground water potential.

Lakes

Krishna district is closely connected with Kolleru lake, which is a large freshwater lake with an extent of about 910 sq. mts. (upto + 10 contour) and an extent of 495 sq. kms. (upto + 5 contour) average. This lake lies between Kaikalur area of Krishna district and Eluru in West Godavari District. One part of the lake lies in Krishna District and the remaining part in West Godavari District. So far as Krishna District is concerned, it lies in Kaikalur and Kalidindi mandals. On the Krishna District side, most of the ponds were excavated within the contours from + 3 to + 5. It acts as a reservoir for drainage water of surrounding delta areas during the monsoons and it dries up in summer.

Literacy

This district is much advanced in education. There are 2317 Elementary Schools, 341 Upper Primary Schools, 277 HighSchools, 33 Junior Colleges, 30 Degree Colleges, 1 Medical College, 3 Engineering Colleges and 2 B. Ed. Training Colleges. There are 12 Welfare Hostels in this district. 53.16 per cent of people are literates in the district as against 45.11 percent of the State average.

Transport and Communication

The district is well served by a network of roads and railways. Nearly 950 villages are connected by Road Transport Corporation buses and the remaining 50 odd villages are connected by private bus routes. Vijayawada city situated at the head of Krishna delta, is the chief commercial centre of the district. It is an important railway junction connected to Delhi, Calcutta, Madras, Hubli, Gudivada and Machilipatnam,. It has an Aerodrome at Gannavaram about 16 Kms. away and a minor seaport at Machilipatnam.

Economy

The main crops grown in the district are paddy, sugarcane, groundnut, chillies, tobacco and cotton. Paddy is grown in Kharif and Rabi seasons as well. Besides paddy the other food crops raised in this district are Jowar and Bajra. Paddy is both wet and dry land crop and the chief food crop in all the mandals. The cry variety is grown on dry lands which do not possess adequate irrigation facilities. The paddy crop on wet land is popular and available. The principal sources of irrigation are river, canals and tanks.

Minerals And Mining

Minerals and mining resources like, Chromite, Iron Ore, Mica and Sulphur are available in considerable quantities. In view of the present demand for Chromite for vizag Steel Plant, it may be worthwhile to carry out detailed exploration in places like, Kondapalli hills and adjoining areas to assess the grade and reserves.

As regards diamonds, a detailed exploration of diamonds, ferrous gravels of Paritala, Ustepalli and other areas is taken up by Geological Survey of India to assess the potentiality of the deposits.

Fishery

The district is quite rich in fishery potential. The main sources of fishery are Krishna river, some part of Kolleru Lake, a number of tributaries like Muniyeru, Budameru, Tammileru and a good number of canals and tanks. The district now has a prestigeous category of freshwater and brackishwater capture and culture fish and shrimp farmers.

Development of Shrimp Farming

The district has a coast line of 88 Kms. The estuarine part of the Krishna river has two main branches with two smaller rivulets in between, which originate from the river and debouch into the sea. In all, there are four such openings of the river into the sea. Besides, there are a number of drains which are influenced by tidal action, as in the case of the four arms mentioned above. Further parallel to the coastline there is narrow backwater formation into which irrigation drains discharge surplus water, mostly from paddy fields. From these water sources certain openings into the sea are formed. These waters are called backwaters which are suitable for shrimp culture.

In addition, the district has 80 sq. kms. area lie hugging the coastline. This area has natural features suitable for the development of brackishwater shrimp farming. Out of the 63, 962 hectares of brackishwater area owned by the State Government all along the coastal belt, 24, 622 hectares (38 per cent of the total area exist in Krishna district. Thus the district has a distinct advantage and enormous potential for brackishwater shrimp farming.

Study Area

The area selected for the present study comprises three villages viz., Kona, Manimeshwaram and Tadivennu situated in the coastal mandals of Bandar, Bantumilli and Kruthivennu respectively. The three sample villages put together have 1356 households belonging to different castes. Out of these 1356 households, 430 are engaged in shrimp culture.

The study area is provided with brackishwater (vacant/forest) lands and drains like Gundlakamma, Pedalanka etc., with necessary tidal fluctuations suitable for shrimp culture. There are primary schools in all the three sample villages. In the mandal headquarters of the three villages there are High Schools and Junior Colleges. In Machilipatnam, the District Headquarters there are three Degree Colleges, one Engineering College and two B. Ed. Training Colleges. MPEDA and BFFDA Offices with qualified staff to assist the shrimp farmers are situated at Machilipatnam.

Indian Bank, State Bank of Hyderabad, Kanakadurga Grameena Bank have their branch offices in the area. The natural shrimp seed is available in good quantities in close vicinity. The agents of shrimp export companies like VBC, Indus Sea Food, Asia Marine Harvest are operating in shrimp marketing.

Socio-Economic Characteristic Features of Sample Farmers

After analysing the profile of the Krishna District and the study area, this chapter examines the socio-economic characteristics of the sample shrimp culture farmers such as caste, literacy, family size and occupational structure and identifies the reasons for undertaking shrimp culture.

Average Size of the Shrimp Farm

The sample respondent farmers are classified into four size groups as presented in the Table 5.1. It is evident from the table that the average size of the farm in terms of acreage varies from 1.46 in marginal farm category to 17.86 acres in big farm category with an overall average of 3.63 acres for he sample as a whole. The data shows that while the average farm size of marginal farmer is 1.46 acres, the average farm size of the small, medium and big farm categories of farmers is 2.75, 7.15 and 17.86 acres respectively.

Family Particulars of Sample Brackishwater Shrimp Culture Farmers

The average size of the family of brackishwater shrimp culture farmers according to the size groups is given in Table 5.2. The total population of the 215 sample households is about 1518 and of this the adult males constitute about 33 per cent, and the adult female and children constitute about 30 per cent and 36 per cent respectively. The sample population is more or less equally distributed as male, female and children. The sample pattern of proportion is not observed in all the size groups. It is found that in

Table 5.1 : Shrimp Culture Pond Area According to Size Group in Sample Villages

Farm Size	*Number of sample farmers*	*Total number of Acres*	*Average size of the Pond in acres*
Marginal	98	144	1.46
Small	82	246	2.75
Medium	20	143	7.15
Big	15	268	17.86
All Farms	250	781	3.63

Table 5.2 : Family Size of Sample Farmers by Size Group

Farm Size	*Family Members*				*Average size of the family*
	Male	*Female*	*Children*	*Total*	
Marginal	236	235	278	749	7.64
	(31.50)	(37.37)	(37.13)	(100.00)	
Small	203	282	222	607	7.40
	(33.44)	(29.98)	(36.57)	(100.00)	
Medium	35	29	31	95	4.75
	(36.84)	(30.53)	(32.63)	(100.00)	
Big	23	23	21	67	4.46
	(34.32)	(34.32)	(31.36)	(100.00)	
All Farms	497	469	552	1,518	7.06
	(32.74)	(30.39)	(36.37)	(100.00)	

Note : Figures in the paranthesis indicate the per centage to the total.

the marginal and small farm size groups, children constitute a high proportion when compared to medium and bigger size groups. The average size of the family is about 7 persons in the case of marginal and small farmers while it is about 5 persons in medium and big farm size groups. Small family norm is yet to be accepted and adopted seriously among the people of all size groups in this area.

Caste Composition

Caste in India is used as a symbol of socio-economic status of the people. Caste barriers in taking up a vocation of a person's own choice proved counter productive in Indian context. People at present are ignoring caste barriers and taking up the professions which in their opinion are profitable and suitable for them.

The participation of different communities among the shrimp cultivators in the three selected villages is shown in Table 5.3. Out of the 215 sample shrimp farmers about 39 per cent belong to Agnikula Kshatriya and Mudiraju communities, the communities which originally belong to traditional fishermen class. The second major community is Scheduled Castes 26.96 per cent, followed by Kapu 16.22 per cent and Kamma 11.16 per cent. Others including Yadavas, Rajaka, Perikala etc., constitute 5.58 per cent and Scheduled Tribes constitute only 0.91 per cent.

The villagewise analysis of communities engaged in shrimp culture reveals that the Agnikula Kshatriyas constitute 75.72 per cent in Tadivennu village of Kruthivennu mandal and their number is Zero in the other two villages, namely Manimeshwaram of Bantumilli mandal and Kona of Bandar mandal. Similarly Mudiraju, another major community engaged in shrimp culture, constitute 42.46 per cent in Kona followed by Scheduled Castes 26.05 per cent, Kammas 19.12 per cent and Kapus 12.32 per cent. In Manimeshwaram Kapus with 34.72 per cent form the major community in shrimp farming, followed by Scheduled Castes 20.85 per cent, Kshatriyas 18.05 per cent and Kammas 13.88 per cent. The data show that while Agnikula Kshatriyas and Mudirajula confine themselves to their respective villages viz., Tadivennu and Kona, Scheduled Caste and Kapu farmers are engaged in shrimp culture in all the three sample villages. Kammas are spread and engaged almost in equal proportion in Manimeshwaram and Kona while their absence was conspicuous in Tadivennu, a fishermen

Table 5.3 : Caste Composition of Sample Shrimp Culture Farmers.

Caste	Manimeshwaram					Tadivennu					Kona					Total				
	M	S	Med	Big	All	M	S	Med	Big	All	M	S	Med	B	All	M	S	Med	B	All
Kamma	–	2	4	4	10	–	–	–	–	0	–	3	3	8	14	–	5	7	12	24
	(14.28)	(40.00)	(66.68)	(13.88)								(15.0)	(33.33)	(100.00)	(14.17)		(609)	(35)	(80.02)	(11.16)
Kapu	24	1	–	–	25	–	–	–	1	1	1	4	4	–	9	25	5	4	1	35
	(57.14)	(7.14)			(34.72)				(100.0)	(1.42)	2.77)	(20.0)	(44.44)		(12.32)	(25.5)	(6.09)	(20)	(6.66)	(16.3)
Kshatriya	1	5	6	1	13	–	–	–	–	0 .	–	–	–	–	–	1	5	6	1	13
	(2.38)	(35.71)	(60.00)	(16.66)	(18.05)											(1.02)	(6.09)	(30)	(6.66)	(6.04)
Agnikula Kshatriya	–	–	–	–	–	15	37	1	–	53	–	–	–	–	–	15	37	1	–	53
						(75.00)	(77.08)	(100.0)		(75.72)						(15.3)	(45.12)	(5)		(24.65)
Mudirajulu	–	–	–	–	–	–	–	–	–	–	16	13	2	–	31	16	13	2	–	31
											(44.44)	(65.00)	(22.23)		(42.46)	(16.36)	(15.85)	(10.00)		(14.41)
S.C.	11	4	–	–	15	1	10	–	–	11	19	–	–	–	19	31	14	–	–	45
	(26.19)	(28.59)			(20.85)	(5.00)	(20.83)			(15.72)	(52.79)				(26.05)	(31.6)	(17.11)			(20.93)
S.T.	–	–	–	–	–	2	–	–	–	2	–	–	–	–	–	2	–	–	–	2
						(10.00)				(2.86)						(2.06)				(0.93)
Others	6	2	–	1	9	2	1	–	–	3	–	–	–	–	–	8	3	–	1	12
	(14.29)	(14.28)		(16.66)	(12.50)	(10.00)	(2.09)			(4.28)						(8.16)	(3.65)		(6.66)	(5.58)

M : Marginal Farmers
S : Small Farmers
Med : Medium Farmers
Big : Big Farmers
Figures in paranthesis indicate Percentages to total

dominant village. Quite significantly, the shrimp culture activity of Kshatriyas is confined only to Manimeshwaram village of Bantumilli mandal, a border mandal of West Godavari District in which Kshatriya is a dominant and enterprising community. It may imply that the Kshatriyas are migrating from nearby villages to Manimeshwaram to engage themselves in shrimp culture.

Literacy Particulars of the Sample Farmers

Literacy is one of the social indicators of development. There is a high correlation between literacy and development. It is generally believed that aquaculture of shrimp on scientific lines would require a certain level of education. It is imperative that brackishwater shrimp farming on scientific and commercial lines would certainly need certain level of literacy, to understand and adopt scientific management techniques. So, to know whether the shrimp farmers had any such education, data are collected on levels of education and presented in Table 5.4.

It can be observed from the table that the majority of the sample farmers, i.e. about 56 per cent are literates, 40 per cent have education upto primary level, 9.30 per cent upto secondary level and 5.11 per cent upto graduate level. Only 1.4 per cent i.e., three in the total sample of 215, have technical education like Engineering, Polytechnique etc. However, in bigger size group the majority of the respondents i.e., 80 per cent, are literate followed by medium size group with 75 per cent, marginal size group with 55 per cent and small farmers group with 48 per cent. The table thus indicates that the majority of the shrimp farmers who have undertaken shrimp farming on scientific lines do have the necessary level of general education. The increase in the literacy level can be attributed to the district administration which is responsible for implementing 'Akshara Krishna' programme effectively in the villages in the recent years.

Occupational Distribution

The sample farmers belonging to various categories were asked about their occupation on the basis of principal and secondary occupation, as occupational structure is often used as the yardstick to measure the level of development.

In the sample of 215 respondents, the majority i.e., about 80

Table 5.4 : Literacy Level of The Sample Farmers by Size Group.

Farm Size	Manimeswaram							Tadivenu							Kona							3 Villages		
	ILLI.	*LIT.*	*P*	*S*	*G*	*T*	*TOT*	*ILLI.*	*LIT.*	*P*	*S*	*G*	*T*	*TOT*	*ILLI.*	*LIT.*	*P*	*S*	*G*	*T*	*TOT*	*ILLI.*	*LIT.*	*TOT*
Marginal	12	30	23	4	3	0	42	11	9	8	1	–	–	20	21	15	11	4	–	–	36	44	54	98
	(28.58)	(71.42)	(54.76)	(9.52)	(7.14)		(100.00)	(55.00)	(45.00)	(40.00)	(5.00)			(100.00)	(58.34)	(41.66)	(30.55)	(11.11)			(100.00)	(44.90)	(55.10)	(100.00)
Small	7	7	4	2	1	0	14	25	23	22	1	0	0	48	11	9	5	2	1	1	20	43	39	82
	(50.00)	(50.00)	(28.51)	(14.28)	(7.15)		(100.00)	(52.09)	(47.91)	(45.83)	(2.08)			(100.00)	(55.00)	(45.00)	25.00)	(10.00)	(5.00)	(500)	(100.00)	(52.44)	(47.56)	(100.00)
Medium	1	9	6	2	1	0	10	–	1	1	–	–	–	1	4	5	2	1	2	0	9	5	15	20
	(10.00)	(90.00)	(60.00)	(20.00)	(10.00)		(100.00)		(100.00)	(100.00)				(100.00)	(44.45)	(55.55)	(22.22)	(11.11)	(22.22)		(100.00)	(25.00)	(75.00)	(100.00)
Big	1	5	2	1	1	1	5	0	1	0	1	0	0	1	2	6	2	1	2	1	8	3	12	15
	(16.67)	(83.33)	(33.33)	(16.66)	(16.66)	(16.66)	(100.00)	(100.00)		(100.00)				(100.00)	(25.00)	(75.00)	(25.00)	(12.50)	(25.00)	(12.50)	(100.00)	(20.00)	(80.00)	(100.00)
All Farms	21	51	35	9	6	1	72	36	34	31	3	–	–	70	38	35	20	8	5	2	73	95	120	215
	(29.17)	(70.83)	(48.61)	(12.50)	(8.33)	(13.8)	(100.00)	(51.42)	(48.57)	(44.30)	(4.27)			(100.00)	(52.06)	(47.94)	(27.39)	(10.95)	(6.84)	(2.73)	(100.00)	(44.19)	(55.81)	(100.00)

P : Primary

S : Secondary

G : Graduates

T : Technical Education

Figures in the Paranthesis indicate percentages to total

Table 5.5 : Shrimp Culture Occupation by Size Group

Farm Size	*Manimeshwaram*			*Tadivennu*			*Kona*			*Total*		
	Principal	*Secondary*	*Total*	*Principal*	*Secondary*	*Total*	*Principal*	*Secondary*	*Total*	*Principal*	*Secondary*	*Total*
Marginal	36	6	42	18	2	20	32	4	36	86	12	98
	(85.72)	(14.28)	(100.00)	(90.00)	(10.00)	(100.00)	(88.88)	(11.12)	(100.00)	(87.75)	(12.25)	(100.00)
Small	10	4	14	44	4	48	17	3	20	71	11	82
	(71.42)	(28.58)	(100.00)	(91.66)	(8.34)	(100.00)	(85.00)	(15.00)	(100.00)	(86.58)	(13.42)	(100.00)
Medium	4	6	10	1	–	1	3	6	9	12	8	20
	(40.00)	(60.00)	(100.00)	(100.00)	–	(100.00)	(33.33)	(66.67)	(100.00)	(60.00)	(40.00)	(100.00)
Big	3	3	6	–	1	1	4	4	8	4	11	15
	(50.00)	(50.00)	(100.00)		(100.00)	(100.00)	(50.00)	(50.00)	(100.00)	(26.66)	(73.34)	(100.00)
All Farms	53	19	72	63	7	70	56	17	73	173	42	215
	(73.81)	(26.39)	(100.00)	(90.00)	(10.00)	(100.00)	(76.71)	(23.29)	(100.00)	(80.46)	(19.54)	(100.00)

Note : Figures in Paranthesis Indicate Percentage to Total.

per cent, reported shrimp culture as their principal occupation. Even according to village-wise classification the majority of the respondents in all the three villages situated in three mandals i.e., (1) Manimeshwaram in Bantumilli Mandal (73.61 per cent), (2) Taidivennu in Kruthivennu Mandal (19 per cent) and (3) Kona in Bandar Mandal (about 77 per cent) have reported shrimp culture as their principal occupation. It is contextual to note that the majority of the sample farmers who undertake shrimp culture as main occupation are from marginal and small farmers. This may imply that these farmers have taken up brackishwater shrimp culture, as most of them are allotted lands by the government and majority of them are fishermen by occupation. Above all, they might have realised that this activity is more remunerative than agriculture or any other alternative activity. The data further show that for majority of the big category of farmers, it is a secondary occupation. It may imply that this category of farmers who have other primary occupation have taken up this venture as secondary occupation as they might have thought that their investment in shrimp farming might give good and quick returns. In the case of medium farmers, shrimp culture is the principal occupation for 60 per cent of the respondents and it is secondary occupation for 40 per cent.

Period of Operation of Sample Farmers by Size Group

In any enterprise, experience matters much for its success and shrimp farming is no exception. The distribution of sample farmers on the basis of the years of experience they had in shrimp culture therefore is analysed in the following page.

Of the 215 sample households selected for the study, 86 farmers (40 per cent)had the experience of 4-5 years (Table 5.6) while 37 farmers (17.20 per cent) had 5 and above years, 26 farmers (12.09 per cent), had 2-3 years and the remaining 26 farmers (12.09 per cent) had 1-2 years of experience in shrimp farming. The data evidently show that more than 50 per cent of the farmers in the study area, on average had more than 4 years of experience in shrimp culture. This justifies the point that the selection of the sample farmers is made on the basis of their experience for a considerable period of time i.e. 4-5 years in shrimp farming. Further, about 19.6 per cent of marginal and small farmers, 26 per cent of medium farmers and 13.33 per cent of big farmers had only 1-2 years of experience. This shows that while marginal and small farmer categories with 14.28 per cent and 21.95 per cent respectively got experience of above 5 years the medium and big farmers size groups constituting 20 per cent and 6.6 per cent respectively had the same experience in shrimp farming. A large per centage of farmers from marginal and small size group had more experience i.e., above 5 years, in shrimp culture than the others. The data also show that 25 per cent of medium and 13.33 per cent of big size group farmers had experience of 1-2 years, while the farmers who had the same experience constitute about 19 per cent in marginal and small farmer size groups. This evidently shows that the farmers belonging to medium and big size groups have been attracted into shrimp culture mostly in . ..ent years.

Table 5.6: Period Of Shrimp Culture Operation of Sample Farmers by Size Group

Farm Size	Number of years of operation					Total
	1—2	2—3	3—4	4—5	5—6	
Marginal	18	14	20	32	14	98
	(18.36)	(14.28)	(20.40)	(32.68)	(14.28)	(100.00)
Small	1	3	15	45	18	82
	(1.22)	(3.67)	(18.29)	(54.87)	(21.95)	(100.00)
Medium	5	5	3	3	4	3
	(25.00)	(25.00)	(15.00)	(15.00)	(20.00)	(100.00)
Big	2	4	2	6	1	15
	(13.33)	(26.68)	(13.33)	(40.00)	(6.66)	(100.00)
All Farms	26	26	40	86	37	215
	(12.09)	(12.09)	(18.60)	(40.00)	(17.22)	(100.00)

Note: Figures in paranthesis indicate percentage to total.

Factors Inducing the Farmers Into Shrimp Culture

Significantly shrimp culture had drawn into its orbit not only the farmers belonging to fishermen community but also people from agriculture and industrial backgrounds. Some farmers had diversified into shrimp culture by converting their agricultural lands into shrimp farming ponds. Diversification of economic activities through shrimp farming is mainly due to the instability and relatively low income generated from agriculture. In addition, a second reason is seepage of saltwater from aquaculture farms to adjacent paddy fields and its adverse affect on paddy crop.

Since shrimp farming is comparatively a new occupation to

many sample farmers, it is quite relevant to know about the factors responsible for their decision to venture into shrimp cultivation. Based on the response given by the sample farmers, various factors have been identified and classified into four broad categories : (1) Availability of suitable land and environment, (2) Profitability, (3) Seepage, (4) Other factors.

Table 5.7 : Factors Inducing The Farmers into Shrimp Culture

Size Group	*Suitability of land*	*Profitability*	*Seapage problem*	*Total*
Marginal	59	18	21	98
	(60.20)	(18.36)	(21.44)	(100.00)
Small	58	13	11	82
	(70.73)	(15.85)	(13.42)	(100.00)
Medium	10	8	2	20
	(50.00)	(40.00)	(10.00)	(100.00)
Big	6	6	3	15
	(40.00)	(40.00)	(20.00)	(100.00)
All Farms	133	45	37	215
	(61.86)	(20.93)	(37.20)	(100.00)

Note : Figures in paranthesis indicate percentage to total.

Availability of Suitable Land and Environment

Primary inducement fro doing shrimp culture is the availability of suitable soil and water. The site should contain soft bottom soil or mixed soil comprising clay, sand and silt to ensure good water bearing capacity as well as production of natural food organism, on which shrimp could feed and grow. The tidal water being the basic water source, should have salinity ranging from 10 to 40 per cent, the ideal salinity for better growth of shrimp is 25 to 35 per cent. The water PH should be 7 to 9. The coastal areas which are subjected to flooding by salt and freshwater during different seasons of the year are found to be having all these properties. The experiments and demonstrations conducted by CIFE have shown that these areas are rich in organic matter and are well suited for the purpose of shrimp culture.

The meterological parameters such as rainfall and its distribution, windspeed and direction, humidity, sunshine and evaporation rate which affect the production are found suitable for the shrimp culture. As per the experiments conducted by CIFE and MPEDA in the study area the levels of materological parameters are ideally suited for shrimp culture operations. In addition, there are a number of drains in this area which are influenced by tidal action. There is not much of mangrove vegetation, ponds can be easily dug with minimum expenditure. Pond sites are also easily accessible. Laboureres are available locally for earth excavation and other civil works in this area. Farming inputs like feed, manures, fertilizers are available in plenty in all the three mandal headquarters.

Brackishwater lands are generally used for forestry, salt manufacturing, creation of sanctuaries, for industrial purposes and tourism development. As there is no specific demand for the brackishwater lands for the above mentioned purposes in the study area, the farmers are using these lands only for shrimp farming purpose. Availability of suitable land and environment therefore is one of the compelling reasons for the sample farmers to plunge into the shrimp farming activity according to the farmers in the study area. About 61.86 per cent of the sample farmers cited that ready availability and suitability of land coupled with other facilities is the main factor that induced them into shrimp farming. Among different size groups, it varies between 60.20 per cent in marginal farmers, 70.73 per cent in small farmers, 50 per cent in medium farmers and 40 per cent in big farmers. The State Government has allotted public lands to the fishermen co-operatives, technocrats and enterpreneurs at 60 : 20 : 20 ratio on realising the use and the demand for these lands. Arrangements are also being made for the distribution of the remaining land at the same ratio in the second phase.

Profitability

Any enterprise depends upon profitability and shrimp farming is no exception. Short duration of crop, quick return on investment and fast expanding attractive world market makes shrimp farming more lucrative and profitable activity .

Aquacultureof brackishwater shrimp is said to be a prime-investment area as it provides wealth and prosperity much more than any other sector. The returns from a unit of aquaculture area of shrimp is expected to be fifteen times higher than the return from one unit of traditional agriculture area. The productivity is also higher in aquaculture i.e, about 25 tonnes per hectare per year, compared to 2 to 5 tonnes per hectare per year in major agricultural crops like cereals. Therefore farmers, irrespective of their caste, occupation and social status, are being induced into this highly profitable venture (Table 5.7). Some of the farmers that belong to the medium and big size groups have even converted their paddy fields into shrimp ponds as they feel that the shrimp farming is more profitable than agriculture.

The table shows that percentage of medium and big sample farmers who have taken up shrimp farming on the profitability criteria is 40 per cent each, whereas it is 18.36 per cent and 13 per cent for the marginal and small farmers respectively. This implies that farmers from the first two categories have taken up shrimp culture as more remunerative enterprise.

Seepage Factor

Based on the above results, it is evident that the main reason for many farmers to take up shrimp farming is suitability of the soil and the next prominent reason is profitability of the soil and the next prominent reason is profitability and the third reason is seepage problem.

Brackishwater aquaculture of shrimp leads to salinisation of land and ground water resources. Salt intrusion takes place due to seepage of pond-held water vertically as well as horizontally. Shallow aquifiers along the aquaculture zone of the coastal belt get salinized. The rate of salinisation depends upon soil characteristics. It is slow in heavy soil with high clay content and its is fast in lighter soils with high sand content. Direct horizontal seepage increases soil salinity directly affecting the crops in the adamant lands. The wells in the coastal villages turn brackish and the water becomes unfit for drinking. In these affected areas shallow tube wells and open wells used for irrigation yield saline water making it difficult for agriculture to prosper. This seepage problem has been identified as one of the factors forcing the farmers to convert their agricultural lands into shrimp ponds. About 17 per cent of total shrimp farmers have expressed that the seepage problem is one of the compelling reasons for them to turn into shrimp culture. As already mentioned, the land subjected to seepage of saline water is totally unfit for paddy cultivation.

Other Factors

In addition to these three identified and highlighted factors, some other factors such as tax exemption given on income obtained through shrimp culture, duty exemption announced for equipment, feed, feed-ingredients, quick return and ready market are also inducing the farmers into the shrimp culture.

Sources of Seed Supply

Seed is an important input for aquaculture of shrimp. The sample respondents in all the three villages are doing monoculture of Tiger shrimp with the exception of only two farmers who are doing polyculture of Tiger shrimp (P. monodon) with white shrimp (P. indicus) under extensive system of cultivation.

Consistent supply of quality seed in sufficient quantities is one of the preconditions for sustained growth of aquaculture of shrimp. In Krishna District the shrimp farmers have two sources to get seed. They are . (1) wild seed and (2) hatchery seed. Most of the farmers depend on wild seed which is a natural is a natural source of seed supply. Table 5.8 presents the sources of shrimp seed. On an average 73.36 per cent of farmers obtain seed from wild source (i.e., from backwaters, coastal waters and estuaries etc.), and 6.97 per cent from hatcheries. About 17.67 per cent comprising all size groups use seed obtained from both the sources. Among different size groups 91.84 per cent of marginal farmers get the seed from a wild source and 4.08 per cent from hatchery, Among small farmers 79.26 per cent get the seed from wild seed stock and a negligible number i.e. less than 2 per cent from hatchery and 19.52 per cent from both the sources. In the case of medium farmers, 20 per cent get the seed from wild source, 25 per cent from hatchery and 55 per cent from both the sources. From among big farmers group only 15 per cent get the seed from wild source, 33.33 per cent from hatchery and 46.57 from both the sources. This shows that medium and big farmer size groups have greater access to hatchery especially the government hatchery TASSPARC of Visakhapatnam where the price of the seed is less when compared to the seed collected and sold through wild source. According to the data a majority of sample farmers get their seed from wild source. This is because there are four important wild seed supply centers in the district namely Konur (Pedapatnam),

Mangenapudi, Karagrahram and Gilakalandindi situated along the coast.

It will not be out of context to mention that natural seed is still of a great magnitude pointing to the fact that we have to traverse a long path to change this pattern.

Table 5.8 : Sources of Shrimp Seed by Size Group

Size Group	*Wild*	*Hatchery*	*Both*	*Total*
Marginal	90	4	4	98
	(91.84)	(4.09)	(4.08)	(100.00)
Small	65	1	16	82
	(79.26)	(2.22)	(19.52)	(100.00)
Medium	4	5	11	20
	(20.00)	(25.00)	(55.00)	(100.00)
Big	3	5	7	15
	(20.00)	(33.33)	(46.67)	(100.00)
All Farms	162	25	38	215
	(75.36)	(6.97)	(17.67)	(100.00)

Note : Figures in paranthesis indicate percentage to total.

Particulars of Land Brought under Shrimp Culture

Indiscriminate conversion of agricultural lands into shrimp culture in the coastal villages have become a subject of criticism at various levels. The environmentalists, backed up by some social service organisations and the local people living in the coastal zones conducted agitations and filed cased in the courts against the persons and firms responsible for the indiscriminate conversion of paddy fields into aquaculture farms. Critics opined that the conversion of certain agriculture lands into shrimp ponds results in fall in food grains productions and create problems of environment pollution and ecological imbalance which ultimately leads to miserable conditions of human life.

Taking the seriousness and significance of this aspect of shrimp culture into consideration, an attempt is made to study this aspect. The lands under brackishwater shrimp culture in the study are drawn from two sources : (1) Vacant/forest lands, (2) Agricultural lands. Out of the total land brought under shrimp culture, vacant or forest lands with brackishwater component constitute a substantially large portion i.e., 78.23 per cent and the agricultural land about 21.77 per cent. On the basis of information given by the sample farmers it is inferred that these vacant and forest lands which have brackishwater component in them are more suitable for shrimp culture. Therefore these lands constitutea higher per centage in the total land under shrimp culture. To ascertain the relative suitability of these lands (vacant/forest lands) over other lands the relevant date are presented in the Table 5.9).

Table 5.9 : Particulars of Land Brought Under Shrimp Culture (In Acres)

Farm Size	*Number of Farmers*	*Vacant/Forest Land*	*Agricultural Land*	*Total Land*
Marginal	98	114	30	144
		(76.25)	(23.75)	(100.00)
Small	82	191	35	226
		(70.43)	(29.57)	(100.00)
Medium	20	93	50	143
		(67.49)	(32.51)	(100.00)
Big	15	213	55	268
		(66.29)	(43.71)	(100.00)
All Farms	215	611	170	781
		(78.23)	(21.77)	(100.00)

Note : Figures in paranthesis indicate percentage to total.

Another interesting result of the data is that the proportion of vacant/forest land, that is brought under shrimp farming by the marginal and small classes of farmers is more i.e., 76.25 per cent and 70.43 per cent respectively, while it is 67.49 and 66.29 per cent in the case of medium and big farmers to the total land under shrimp culture. The marginal and small farmers have possessed more of vacant and forest lands, because most of these lands are public lands, allotted to them by the government.

The proportion of agricultural land brought into shrimp culture is less in the case of marginal and small farmer is.e., 23.75 per cent and 29.57 per cent respectively, and more in the case of medium and big farmers, i.e., 32.51 and 43.71. Farmers irrespective of farm size converted their agricultural lands which are adjacent to shrimp culture ponds due to the problem of seepage of saltwater into paddy fields. Farmers belonging to medium and big firm categories are originally agriculturists. Some of them are business people belonging to the nearby towns. They have entered into this venture and converted their paddy growing lands into shrimp culture ponds as they found shrimp culture more profitable than agriculture. The data reveals further, that the last four or five years of shrimp culture activity in the study area, could result in conversion of only about 22 per cent of agricultural land into shrimp farming. The other 78 per cent of land under shrimp culture is either derelict vacant land or a forest land.

Conclusion

Fisheries, which is considered as caste based 'low class occupation' has suddenly turned into a high class enterprise with no caste constraint. Farmers belonging to various caste groups are drawn into shrimp culture without any caste barriers because they found this activity most profitable and lucrative.

The major factors that are inducing the farmers into brackishwater shrimp culture are identified as : (1) Availability of suitable land and environment; (2) Profitability; (3) Seepage factor; and (4) Other factors. The study locality is situated along the coast line spread across the three coastal mandals with conducive land properties and congenial environmental factors for shrimp farming. The majority of the farmers got considerable experience in shrimp culture and education upto primary level.

From the distribution of the heads of sample households, it is evident that the farmers belonging to various castes and sizes of land are involved in extensive type shrimp culture. The brackishwater lands which were considered 'derelicit' became 'prime lands'. The coastal farmers who lacked alertness and ambition have suddenly become active and enterprising with the shrimp culture activity.

Chapter 6

Costs and Returns

Introduction

The profitability and economic viability of shrimp farming, as in any other enterprise, depends upon the input-output ratio *i.e.* the relationship between the cost incurred on inputs and the income obtained from output. Profit maximisation depends upon the reduction in the cost of production. Reducing costs and increasing income requires the entrepreneur to make judicious development and rational utilisation of inputs. This chapter is organised into two sections. Section-I deals with costs and returns in brackishwater shrimp farming, while Section-II provides the production function analysis (input-output relationship).

SECTION-I

Non-Recurring (Capital) Investment in Shrimp Culture

Prior to the analysis of the costs and returns, we may examine the nature of capital investment in shrimp culture. Like any other commercial venture, shrimp farming too, requires investment both of non-recurring and recurring nature. As the pattern of investment is one of the most crucial determinants of productivity and profitability we may have to look into various aspects of investment. Non-recurring capital investment in shrimp farming covers excavation of the pond area, construction of main and subsidiary bunds, sluice gates, construction of watchmen shed-cum store house, cost of equipment such as pump sets, aerators, transport conveyances, etc. These components may vary depending upon the size of the farm and its nature of operations. Details of non-recurring capital among the sample farmers are worked out according to the size of the farm and are presented in Table 6.1. The table shows an average non-recurring capital investment of Rs. 17,667 per acre, for the sample as a whole. The same investment per acre varies between Rs. 12,989 and Rs. 27,988 from marginal farms to big farms. Further, the farm size and investment per acre are directly related. This is because of costly equipment purchased, construction of pucca sheds, and vehicles used by the farmers with the increase in the farm size.

As to the composition of total non-recurring capital investment of all the sample farmers put together, the proportion of investment on digging to the total non-recurring investment accounts for about 61 per cent, followed by equipment for 22.58 per cent and sheds for 16.52 per cent .

Table 6.1 : Non-Recurring Capital by Item and Size Group per Acre (In Rupees)

Farm Size	*Digging and inlet-outlet construction*	*Buildings/Sheds and Store Room*	*Equipment*	*Total*
Marginal	11,150	835	1,004	12,989
	(85.84)	(6.43)	(7.73)	(100.00)
Small	10.450	1,245	2,755	14,450
	(72.32)	(8.62)	(19.06)	(100.00)
Medium	9,925	3,166	4,309	17,489
	(56.74)	(18.10)	(25.14)	(100.00)
Big	9,835	5,979	7,174	22,988
	(42.80)	(26.00)	(31.20)	(100.00)
All Farms	10,271.90	3,145.60	4,249.36	17,666.86
	(58.14)	(17.80)	(14.05)	(100.00)

Note : Figures in paranthesis indicate the percentage to the total.

Digging and inlet-outlet construction cost decreases as the size of the farm increases, which is 85.84 per cent in the case of marginal farms and 42.80 per cent for big farms. On sheds and store rooms it is the lowest *i.e.,* 6.43 per cent in the case of marginal farms and the highest (26 per cent) in the case of big farms. The differences in the percentage of investment on sheds and buildings between the marginal and big farms is due to the differences in size and the type of sheds and buildings constructed by these farmers. The size of the farm and the investment on sheds and buildings are directly related.

In the case of equipment the proportion of investment is found to be highest on big farms (31.20 per cent) and lowest (7.73 per cent) on marginal farms. The size of the farm and the expenditure on equipment are directly related. This difference may be related to the utilisation of modern equipment in the case of big and medium farms and lack of economic viability in the case of marginal and small farms.

Recurring (Variable) Costs in Shrimp Culture

The operational costs of recurring nature are called recurring costs or working costs. These costs vary with the level of production. In shrimp culture the recurring costs include cost on pond preparation, seed, feed, fertilisers, fuel, repairs and replacements of bunds and equipment, wages for workers and other miscellaneous expenses.

Table 6.2 : Recurring Cost of Shrimp Culture According to Cost Items by Size Group Per Acre 9 in (Rupees) (Per Two Crops)

Size Group	*Seed*	*Feed*	*Labour*	*Fuel*	*Miscellaneous*	*Total*
Marginal	8,395	9,445	4,878	1,700	2,920	25,338
	(30.70)	(33.20)	(17.84)	(6.72)	(11.54)	(100.00)
Small	8,845	9,628	5,230	2,001	2,746	28,450
	(31.08)	(33.84)	(18.38)	(7.03)	(8.67)	(100.00)
Medium	8,236	11,186	5,954	2,145	2,660	30,191
	(27.28)	(37.05)	(19.75)	(7.10)	(8.82)	(100.00)
Big	8,842	12,325	6,023	2,468	2,543	32,201
	(27.45)	(38.27)	(18.70)	(7.66)	(7.89)	(100.00)
All Farms	8,649.49	10,805.00	5,571.61	2,132.11	2,692.67	29.850.61
	(28.97)	(36.19)	(18.66)	(7.14)	(9.02)	(100.00)

Note : Figures in the paranthesis indicate the per centage to the total.

The details of the recurring cost of production per Kg. and cost of cultivation per acre for two crops is shown in Table 6.2 by size group. As the figures in the table show, the recurring cost of shrimp per acre varies between Rs. 27,338 on marginal farms to Rs. 32,201 on big farms. For he sample as a whole it is worked out at Rs. 29,850.61. Direct relationship is observed between the size of the farm and the recurring cost. The recurring cost increases with the increase in the size of the farm.

Production Costs in Shrimp Culture

The total cost of production consist of fixed and variable costs of production. Fixed costs are those costs which are independent of the level of production. These costs do not change with the change in the size of production. Fixed costs include depreciation on equipment and interest on fixed capital. The variable cost includes the cost incurred on seed, feed, labour, fertilisers and other miscellaneous expenses. The details of the production cost per acre and per Kg. are shown in Table 6.3.

Table 6.3. : Production Cost in Shrimp Culture per Acre by Size Group (In Rupees)

	Fixed Costs		*Variable Costs*		*Total Cost*	
Farm Size	*Per Acre*	*Per Kg.*	*Per Acre*	*Per Kg.*	*Per Acre*	*Per Kg.*
Marginal	1,482.90	0.09	27,338	167.71	28,821	176.80
	(5.15)		(94.85)		(100.00)	
Small	3,001.00	17.55	28,450	166.36	31,451	183.91
	(9.55)		(90.45)		(100.00)	
Medium	3,904.40	20.54	30,191	158.90	34,095	179.44
	(11.45)		(88.55)		(100.00)	
Big	5,453.00	26.60	32,201	157.07	37,654	183.67
	(14.48)		(85.52)		(100.00)	
All Farms	3,919.00	20.26	29,850.61	161.84	33,770	182.10
	(11.60)		(88.40)		(100.00)	

Note : Figures in paranthesis indicate per centage.

The table shows, that for the sample as a whole, the total cost of production is Rs. 33,770 and of this variable costs alone constitute about 88 per cent. Thus it is evident, that the total cost increases with the size of the farm from Rs. 28,821 on marginal farms to Rs. 37,654 on big farms. There is direct relationship with regard to the variable cost and the size of the farm.

The proportion of fixed cost to the total production cost is highest on big farms i.e. 14.5 per cent and lowest on marginal farms, about 5.15 per cent. The table shows that the size of the farm and fixed cost are also directly related. The fixed cost per acre increases with the increase in the size of the farm. However the proportion of variable cost to the total cost of shrimp is found to be inversely related. The proportion of variable cost to the total cost is highest on marginal farms i.e. about 94.85 per cent and lowest on big farms, about 85.52 per cent.

To obviate the effect of sheer farm size on cost of production, comparison is made with respect to the cost of production per Kg. of shrimp. It is pointed out that the production costs per Kg. of cultured shrimp exhibit slight variation among different size group of farms. It is Rs. 179.44 on medium farms and Rs. 183.67 on big farms. The average cost of production per Kg. of shrimp is estimated at Rs. 182.10. Though the variation in cost of production is not significant between various size groups, it is high in small farms and low in marginal farms. The lack of significant variation in the cost of production between various size groups can be attributed to the scientific nature of brackishwater shrimp farming. Farmers irrespective of their farm size use inputs almost in the same proportion. It may imply that the differences in cost of production per acre between various size groups is mainly due to the differences in the fixed costs and to some extent due to the differences in variable costs.

Production Costs According to Cost Items

To assess the relative importance of different items of total cost, we have to examine the components of various cost items, which include, among others, cost of seed, feed, labour charges including the imputed value of family labour, fuel charges, other miscellaneous expenses, interest on fixed capital and depreciation of fixed assets. Details relating to the cost items incurred by the sample shrimp farmers are presented in Table 6.4.

The table shows, that out of the total cost of production, feed cost alone constitutes 32 per cent and among different size groups it varies between 30 and 32.80 per cent *i.e.*, 30.60 per cent on small farms to about 32.75 per cent on medium and big farms. This small variation is due to the scientific nature of this culture. Lack of significant difference in the fixed cost can be attributed to the lack of differences in the quantity and quality of feed used by the sample farmers irrespective of their size group. Therefore definite conclusions cannot be made about the relationship between feed cost and farm size.

Next to feed, expenses on seed constitute a major share of the total cost, which accounts for 25.61 per cent of the total cost for the sample as a whole. It ranges from 28.48 per cent for the big farm group to 29.13 per cent for marginal farm group indicating

Table 6.4 : Production Cost of Shrimp Culture According to Cost Item per Acre by Size Group (In Rupees)

Items	*Marginal*		*Small*		*Medium*		*Big*		*All Farms*	
	Per Acre	*Per cent to total*	*Per Acre*	*Per cent to total*	*Per Acre*	*Per cent to total*	*Per Acre*	*Per cent to total*	*per Acre*	*Per cent to total*
Seed	8,395	29.13	8,845	28.12	8,236	24.15	8,842	23.48	8,649.49	25.61
Feed	9,445	32.80	9,628	30.60	11,186	32.80	12,325	32.73	10,805.00	32.00
Labour	4,878	17.00	5,230	16.61	5,954	17.50	6,023	16.00	5,571.61	16.50
Fuel	1,700	6.00	2,001	6.39	2,145	6.30	2,468	6.55	2,132.11	6.32
Miscellaneous	2,920	10.13	2,746	8.73	2,660	7.80	2,543	6.76	2,692.67	7.97
Total Variable Cost	27,338	94.85	28,450	90.45	30,191	88.55	3,220	85.52	29,850.61	88.40
Fixed Cost :										
Depreciation	183.90	0.64	400.00	1.27	756.4	2.22	1,315	3.50	739	2.18
Interest	1,298.90	4.51	2,601.00	8.28	3,148.6	9.23	4,138	10.98	3,180	9.42
Total Cost	1,482.80	5.15	3,001.00	9.55	3,905.00	11.45	5,453	14.48	3,919	11.60

thereby a difference of 5.65 per cent between the two groups. It indicates that as the size of the farm increases, the seed cost decreases. This is mainly due to the difference in the sources of seed they purchase. The big farmers generally buy the seed from the government hatcheries where the price is less. The marginal and small farmers on the other hand buy the seed procured through natural sources and sold through middle men at a relatively higher price than the hatchery produced seed.

Expenditure incurred on labour constitutes 16.50 per cent of the total cost for the sample as a whole and it ranges from 16 per cent on big farms and about 17 per cent on marginal and medium farms, indicating no significant difference between various size groups. Expenditure on miscellaneous and other charges is about 8 per cent of the total cost for the sample as a whole and the percentage between various size groups varies between 10.13 per cent on marginal farms and 6.76 per cent on big farms, indicating an inverse relationship between the size of the farm and the miscellaneous expenditure incurred. As the size of the fram increases the expenditure on miscellaneous items decreases. This may be due to the economies of large scale production. The expenditure on fuel is worked out at 6.32 per cent for the sample as a whole and the percentage is almost the same on different size groups.

Among fixed items, interest constitutes 9.42 per cent of the total cost and it varies between 4.51 per cent on marginal farms and 10.98 per cent on big farms. It shows that the share of interest in the total cost increases as the size of the farm increases. It may imply that as the amounts of money borrowed by medium and big farmers is more, their per cent of interest in the total cost of production is also more. Depreciation constitutes a negligible 2.18 per cent in the total cost for the sample as a whole and ranges from 0.64 per cent on small farms to 3.50 on big farms, indicating thereby a difference of about 3 per cent. This marginal difference in depreciation charges is due to the differences in the equipment used by different size groups.

Shrimp Production

The details of total production of shrimp according to farm size is collected from the farmers belonging to various size groups on the basis of oral information given by them. Shrimp production, according to farm size is given in Table 6.5. The total cultured shrimp production from all sample farmers is 1,51,068 Kilograms and the average production per acre is 193.42 Kgs. Among different size groups, the shrimp production per acre varies between 163 Kgs. on marginal farms and 205 kgs. on big farms per acre. As per the table, a direct relationship is observed between the size of the farm and production. As the size of the farm increases, the shrimp production also increases. This may perhaps be an indication of the economies of large scale production.

Table 6.5 : Shrimp Production by Size Group

Farm Size	*Production of shrimp (Kilograms)*	
	Total Production	*Per Acre*
Marginal	22,472	163
Small	45,486	171
Medium	27,170	190
Big	54,940	205
All Farms	1,51,068	193.42

Net Returns in Shrimp Culture

Net economic return in shrimp culture depends upon the production cost. As mentioned already, the production cost includes the total recurring (variable) costs and the total fixed costs. The net returns per acre, per year (two crops) for each size group are obtained by deducting the production cost from the gross returns.

As per the Table 6.6 the overall net returns per acre of shrimp is worked out at Rs. 50,008, ranging from Rs. 43,714 on marginal farms to Rs. 56,641 on big farms, the average net return per Kg. being Rs. 258.55 for the sample as a whole. It is varied between Rs. 268.18 on marginal farms to Rs. 275.32 on big farms.

As regards the average net return per acre, a direct relationship is observed between the returns obtained and the size of the farm. The net returns increase with the increase in the size of the farm. But as regards the output-input ration, no direct relationship is noticed. It is 2.52 in the case of marginal farms and 1.50 for big farms. The output-input ratio varied between 2.52 on marginal farm group and 1.50 on big farm group, with an overall average of 1.48 for the sample as a whole.

Table 6.6 : Net Returns in Shrimp Culture : By Size Group

(In Rupees)

Farm Size	*Production Cost*		*Gross Returns*		*Net Returns*		*Output-Input Ratio*
	Per Acre	*Per Kg.*	*Per Acre*	*Per Kg.*	*Per Acre*	*Per Kg.*	
Marginal	28,821	176.81	72,535	445	43,714	268.18	2.52
Small	31,450	183.91	76,779	449	45,329	265.08	1.44
Medium	34,095	179.45	86,830	457	52,735	277.55	1.55
Big	37,654	183.68	94,095	459	56,441	275.32	1.50
All Farms	33,770	174.59	83,779	433	50,008	258.55	1.48

Profitability of Shrimp Culture

For assessing profitability, we have to examine : (i) the rate of net returns and (ii) the returns on investment in shrimp culture. To arrive at the rate of returns, gross returns has been considered and to calculate returns on investment, the capital investment and the production cost have been taken into account.

Rate of Return

The rate of return in shrimp culture is obtained by the percentage of net return in the gross returns which is presented in Table 6.7. No significant difference is observed among different farm groups in he rate of returns. The rate of return is found to be highest in the medium farm group with 60.73 per cent, followed by big farm group with 59.98 per cent, marginal farm group with 59.85 per cent and small farm group comes next with 59.03 per cent. No direct relationship is observed between the rate of return and size of the farm.

The operational efficiency is generally judged on the basis of the rate of return *i.e.*, net profit obtained per acre or income earned in relation to investment.

Table 6.7 : Rate of Net Return In Shrimp Culture per Acre : By Size Group

Farm Size	*Net Returns*	*Production Per Acre (in Kgs.)*	*Sales Return Per Acre*	*Rate of Return (%)*	*Average price*
Marginal	43,714	163	72,535	59.85	445.00
Small	45,329	171	76,779	59.03	499.00
Medium	52,735	190	86,830	60.73	457.00
Big	56,441	205	94,095	59.98	459.00
All Farms	50,008	193.42	83,779	59.69	433.12

These indicators are generally influenced not only by the inputs used but also by the experience and the knowledge of the farmers in pond management, disease identification etc. There is no direct relationship between the rate of return and the size of the farm. No significant difference is observed in the rate of return accrued to different size groups of farms. Absence of any such noticeable difference in the rate of returns may be due to the following reasons.

1. Brackishwater shrimp culture is scientific in its nature. Therefore the farmers irrespective of their economic viability and farm size, use various inputs for a unit area as per the specifications and guidelines of the MPEDA.

2. As brackishwater shrimp culture is considered as hundred per cent export oriented venture, the farmers, irrespective of their farm size, are able to secure necessary loans and other inputs without much difficulty. Therefore, not much difference is observed between the farmers of various size group in the quantity and quality of inputs and the rate of returns obtained by them. The marginal difference observed in the rate of return received by various size groups may be perhaps due to the differences in the operational efficiency of individual farmers and other environmental advantages that their farms had in the production year.

Returns on Investment

Returns on investment in brackishwater shrimp culture is obtained by expressing net returns as percentage of the total investment and the details are presented in Table 6.8.

Table 6.8 : Net Return on Investment in Shrimp Culture per Acre by Size Group

					Percentage net return over investment	
Firm Size	*Capital Investment*	*Production Cost*	*Total Investment*	*Net Return*	*With Production Cost*	*Without Production Cost*
Marginal	12,989	28,821	41,810	43,714	109.55	336.54
Small	14,450	31,450	45,900	45,329	98.75	313.69
Medium	17,489	34,095	51,584	52,735	102.23	301.53
Big	22,988	37,654	60,642	56,441	93.07	245.52
All Farms	17,669	33,770	51,437	50,008	97.22	283.05

The table indicates that the returns on investment in shrimp culture seem to be very encouraging under the conditions present in the study area. For the sample as a whole, the net return on investment for the entire production period (about 9 months for two crops) is about 97 per cent. When the size of the farm is taken into account, it ranges from 104.55 per cent on marginal farms to about 93 per cent for the big farms. However, the net returns on investment did not indicate any relationship between the size of the farm and the net returns on investment. The figures show, that in regard to net returns on investment,the marginal farm is in an advantageous position. This is perhaps due to the less cost of production incurred by the marginal farmers per unit area *i.e.,* one acre than that incurred by the farmers belonging to the other size groups and other favourable conditions that this size group enjoyed during the year of study.

From the results drawn from the study, it is found that brackishwater shrimp culture is certainly the most profitable, lucrative and economically viable enterprise among all the alternative allied activities. The sudden shift of farmers towards this culture and the rush of private entrepreneurs including some corporate houses into this new venture, therefore is quite understandable.

SECTION-II

Production Function

So far the costs and returns of shrimp culture have been analysed. Now we have to examine the factors affecting shrimp production by fitting production function. Generally, a production function expresses the technical relationship between output or production of a commodity and inputs or factors of production . In a study conducted by Dhondyal and Singh,[1] a production function was specified with a few estimated variables viz., labour charges per acre and feed charges per acre and the dependent variables being output of fish per acre in rupees. Though they estimated the production function, they did not test the significance

1. Dhondyal, S.P., and O.N. Singh, 'Cost benefit ratios in fish culture', *The Indian Journal of Agricultural Economics,* Vol. XXIII, No. 4, Oct.-Dec., 1986, pp. 235-239.

of the model. Singh and Pandey[2] while estimating production function in their study they have considered a single explanatory variable, *i.e.*, input tackle as a sole factor explaining variations in fish production.

Keeping the above studies in view, in the present study, shrimp production (measured in Kgs.) during the study year is considered as the dependent variable. The factors of production or the explanatory variables include seed (number of fingerlings), inorganic fertilizer (Kg.), supplementary feed (Kg.), human labour (mondays) and the cost of other miscellaneous inputs. The production function equation to be estimated is :

$$Q = f\,(S, IF, SF, HL, M)$$

Assuming a linear relationship between the shrimp production and explanatory variables, the production function can be written as :

$$Q = b_o + b_1S + b_2IF + b_3SF + b_4HL + b_5M + U$$

where Q = Shrimp production (Kgs.) per acre of pond area;

S = Seed (no. of fingerlings) per acre of pond area;

IF = Inorganic fertilizer (Kgs.) per acre of pond area;

SF = Supplementary feed (Kgs.) per acre of pond area;

HL = Human labour employed (Mandays) per acre of pond area.

M = Miscellaneous input charges (Rs.) per acre of pond area which include preparatory expenses, water charges, expenses on repairs, tahasil etc.

b_o = Constant to be estimated;

b_1 = Regression coefficient of the 'i' the variable to be estimated (where i = 1 to 5); and

U = Random error.

Ordinary Least Squares method has been employed to estimate the regression parameters. The regression equation is estimated for all farms category as well as for different size g roups of farms.

---

2. Singh, I.J. and U.K. Pandy, 'Fish Production Function for West Godavari District, (A. P.)', *The Indian Journal of Agricultural Economics*, Vol. XXIII, No. 4, Oct.-Dec., 1986, pp. 255-258.

Results

Prior to the interpretation of the estimated regression parameters the severity of the problem of multi-collinearity has been tested with the help of zero-order correlation matrix and found that there is no severe multi-collinearity among the explanatory variables.[3]

All Farms

The zero order correlation matrix is presented in Table 6.9. It indicates the absence of multi-collinearity in the estimated regression equation. The estimated regression equation is :

$$Q = 60.18 \underset{(0.0044)}{\overset{*}{+ 0.0106S}} \underset{(0.4251)}{+ 0.0880IF} \underset{(0.0355)}{\overset{*}{+ 0.0904SF}} \underset{(0.0060)}{- 0.0067HL}$$

$$\underset{(0.0052)}{\overset{*}{+ 0.0129M}}$$

$\overline{R}^2 = 78.62$ PC.; F Value = 127.47

All the explanatory variables put together explain about 78 per cent of the variation in shrimp production and this coefficient of multiple determination is found to be significant at 1 per cent probability level. The regression coefficients associated with all the explanatory variables except human labour are positive. Only the coefficient associated with human labour registered a negative sign. However, this coefficient is found to be not significant at 10 per cent probability level. Among other coefficients, the coefficients associated with seed, supplementary feed and miscellaneous expenses are found to be significant at 5 per cent probability level, while the coefficients associated with inorganic fertilizers is insignificant even at 10 per cent probability level. The estimated

---

3. According to Heady, if the correlation coefficient between the explanatory variables is less than 0.8 then it can be considered that there is no multi-collinearity. See Heady, Earl O., and j.L. Dellion, *Agricultural Production Functions*, Iowa State University Press, Ames, Iowa, 1964.

* Indicates significance at 5 per cent level.

** Indicates significance at 1 per cent level.

coefficients indicate that a kilogram increase in the use of supplementary feed and inorganic fertilizers will increase shrimp production by 0.0904 and 0.088 kilograms respectively.

Table 6.9 : Zero Order Correlation Matrix

Farm Size	*Q*	*S*	*IF*	*SF*	*HL*	*M*
All Farms						
Q	1.00	0.18	0.37	0.59	-0.20	0.09
S		1.00	-0.07	-0.28	0.54	-0.07
IF			1.00	0.39	-0.13	0.01
SF				1.00	-0.24	0.90
HL					1.00	0.01
M						1.00

Marginal Farms

The zero order correlation matric of marginal farms is presented in Table 6.10. As per the table the estimated correlation coefficients indicate the prevalence of multi-collinearity among some of the explanatory variables. However, none of the correlation coefficients exceed 0.80 and hence there is not much multi-collinearity (Heady). The estimated regression equation is :

$$Q = 52.49 \underset{(0.0130)}{+ 0.0133S} \underset{(0.0363)}{+ 0.08IF^{*}} \underset{(0.0203)}{+ 0.0666SF^{**}} \underset{(0.0217)}{- 0.0133HL} \underset{(0.0168)}{+ 0.008IM}$$

$\overline{R}^2 = 56.29$; F Value = 19.53

In case of marginal farms, the coefficient of multiple determination is 0.56 and it is found to be significant at 1 per cent probability level. This implies that the explanatory variables considered in the model explain about 56 per cent of the variation in shrimp production. Only the coefficient associated with the human labour has a negative sign. Coefficient for supplementary feed is significant at 1 per cent probability level while that for inorganic fertilizers it is insignificant at 5 per cent probability level. The coefficients associated with seed, human labour and miscellaneous costs are not significant at 5 per cent probability level. The estimated coefficients indicate that an additional use of one kilogram per acre of inorganic fertilizer and supplementary feed will increase shrimp production by about 0.08 and 0.066 kg. respectively. Similarly, an increase in the seed by one fingerling will enhance shrimp production by about 0.0133 kg.

Table 6.10 : Zero Order Correlation Matrix

Farm Size	*Q*	*S*	*IF*	*SF*	*HL*	*M*
Marginal Farms						
Q	1.00	0.09	0.53	0.43	-0.02	0.04
S		1.00	0.05	- 0.01	0.75	0.22
IF			1.00	0.48	- 0.05	0.31
SF				1.00	-0.04	0.18
HL					1.00	0.20
M						1.00

Small Farms

The zero order correlation matrix of the small farm category is presented in Table 6.11 and these coefficients indicate the absence of multi-collinearity in the data. The estimated regression equation is :

$$Q = 50.10 \underset{(0.0024)}{+ 0.0147S^{**}} \underset{(0.0365)}{+ 0.0399IF} \underset{(0.0189)}{+ 0.0532SF^{**}} \underset{(0.0156)}{- 0.0081HL} \underset{(0.0074)}{+ 0.0107M}$$

$$\bar{R}^2 = 50.88; \qquad F\ Value = 12.94$$

Table 6.11 : Zero Order Correlation Matrix

Farm Size	*Q*	*S*	*IF*	*SF*	*HL*	*M*
Small Farms						
Q	1.00	0.01	0.09	0.52	0.14	0.01
S		1.00	0.09	0.27	0.29	0.34
IF			1.00	0.20	0.03	-0.08
SF				1.00	0.07	-0.07
HL					1.00	0.48
M						1.00

All the explanatory variables put together explain about 50 per cent of the variation in shrimp production and all the regression coefficients with the exception of human labour turn to be positive.

Though the coefficient associated with human labour is found to be negative it is not significant even at 10 per cent probability level. In this category of farms, only the coefficients associated with seed and supplementary feed are found to be statistically significant at a probability level of 1 per cent. The results further indicate that an increase in the use of seed by one fingerling will increase shrimp production by about 0.0147Kg. Further, an additional use of inorganic fertilizer and supplementary feed by 1 Kg. per acre will increase shrimp production by about 0.0399 and 0.0532 Kg. respectively.

Medium Farms

The estimated zero order correlation matrix of medium farm category is presented in Table 6.12 and these coefficients reveal the absence of multi-collinearity in the data. The estimated regression equation is :

Table 6.12 : Zero Order Correlation Matrix

Farm Size	*Q*	*S*	*IF*	*SF*	*HL*	*M*
Medium Farms						
Q	1.00	0.26	-0.61	0.65	- 0.07	0.05
S		1.00	0.07	0.24	0.20	0.02
IF			1.00	0.45	0.01	-0.11
SF				1.00	0.11	0.03
HL					1.00	0.10
M						1.00

$$Q = 61.69 \quad \underset{(0.0151}{+0.0160S} \quad \underset{(0.0389)}{\overset{*}{+0.0846IF}} \quad \underset{(0.2511)}{+0.2356SF} \quad \underset{(0.0240)}{-0.0071HL}$$

$$\underset{(0.0032)}{\overset{**}{+0.0102M}}$$

$\bar{R}^2 = 81;$ F Value = 9.23

The coefficient of multiple determination of the medium farm function is about 0.81 and it is found to be significant at 1 per cent probability level. This indicates that in the selected model all the explanatory variables put together explain about 81 per cent of variation in the dependent variable i.e., shrimp production. The coefficient associated with miscellaneous expenses is found to be significant at 1 per cent probability level. In this category the coefficient for inorganic fertilizers and for human labour registered a negative sign. But the coefficient for human labour it is not significant even at 10 per cent probability while it is significant for inorganic fertilizers at 5 per cent probability level.

Big Farms

The estimated zero order correlation matrix of big farms is presented in Table 6.13.

Table 6.13 : Zero Order Correlation Matrix

Farm Size	*Q*	*S*	*IF*	*SF*	*HL*	*M*
Big Farms						
Q	1.00	0.04	- 0.11	0.20	0.36	- 0.40
S		1.00	0.26	- 0.01	- 0.28	- 0.16
IF			1.00	0.18	- 0.19	- 0.29
SF				1.00	0.36	-0.14
HL					1.00	0.28
M						1.00

The estimated regression equation is :

$$Q = 88.15 \; + \underset{(0.0029)}{0.0102S^{**}} \; - \underset{(0.0425)}{0.0986IF^{*}} \; + \underset{(0.3486)}{0.2912SF^{**}} \; + \underset{(0.0765)}{0.2282HL}$$

$$+ \underset{(0.0020)}{0.0082M^{**}}$$

$\bar{R}^2 = 83$; F Value = 6.51

In this category of farms the coefficient of multiple determination is found to be 0.83. This means that the explanatory power of the model is about 83 per cent. The coefficients associated with seed and human labour are positive and are found to be statistically significant at 1 per cent probability level. The coefficients associated with miscellaneous expenses and inorganic fertilizers are negative and are significant at 1 per cent and 5 per cent probability levels respectively.

The results of the estimated production function reveal that supplementary feed, seed and miscellaneous expenses are the important variables explaining variation in shrimp production.

Conclusion

The rush of private entrepreneurs into brackishwater shrimp culture, and the sudden shift of farmers from agriculture to aquaculture is said to be due to the profitability of shrimp farming. The costs and returns aspect of this study strengthens this point. The overall net return per acre of shrimp is worked out at Rs. 50,008 ranging from Rs. 43,714 on marginal farmers to Rs. 56,441 on big farmers. The average rate of return is found to be about 60 per cent for all the size groups. Though the cost of production and net returns increase with the increase in the size of the farm, no direct relationship is observed between the rate of return and the size of the farm. Lack of significant difference in the rate of returns may be attributed to the scientific nature of shrimp culture where farmers irrespective of their economic viability and farm size use the same quantity of inputs per one unit area. From the results drawn through this study it is clear that the brackishwater shrimp culture is certainly more profitable and lucrative.

The results of the estimated production function reveal that supplementary feed, seed and miscellaneous expenses are the important variables explaining variations in shrimp farming.

Chapter 7

Employment in Brackishwater Shrimp Culture and Market Structure

Introduction

This chapter is divided into two sections. Section-I explains the employment generation in brackishwater shrimp culture and Section-II examines the market structure of brackishwater shrimp culture in the study area.

SECTION-I

Employment in Brackishwater Shrimp Culture

"Jobs rather than men, should wait" said Sir William Beveridge.1 But the situation in an over populous and a developing country like India is entirely different. Unemployment problem of unprecedented nature persists at present in Indian economy. India's planning strategy for its labour surplus economy therefore has been an employment oriented growth strategy. In the agro-based economy like India the rural poor consists of landless labour besides small and marginal farmers. The redundant rural labour could be absorbed in gainful occupations only through diversification of labour use. Although a good number of programmes have been launched by the government from time to time in this regard, the unemployment scenario in rural India by and large remains the same. It is in this context that the brackishwater aquaculture of shrimp is found to be prominent as a potential source of income and employment generation to the rural poor.

1. Ruddar Dutt and Sundaram, K.P.M., *Indian Economy*, S. Chand & Company Ltd., New Delhi, 1995, p. 343.

Brackishwater shrimp culture offers direct employment in such activities as digging of ponds, cleaning of ponds, fertilisation, seeding, feeding, water management, trialnetting and final netting, supervision, packing, marketing etc. Besides this, development and expansion of shrimp culture also promotes indirect employment in industries producing feed and fertilisers, transporting of feed and seed, collection of wildseed, snails, basket making, ice supplying, processing etc. Further, the scientific and commercial shrimp culture involves construction of houses for supervisors, storerooms, pump sheds, dykes, feeder canals, water reserve tanks etc., which generates employment and creates demand for, semi-skilled and unskilled rural labour and enhances their income.

The labour obsorptive capacity of shrimp culture can be noticed in the magnitude of human labour employment by examining the shrimp farming in the study area with that view point.

In the study, employment of human labour in brackishwater shrimp culture is worked out for one production year (two crops) on the basis of actual hours of work done by family as well as hired labour, both male and female, at the farm level. For the purpose of computing labour absorption rates, work done by labour is expressed in manday equivalent. The manday is defined narrowly as equivalent to an eight hour work done by either male or female. No distinction is made between the work of males and females as the nature of work is such that it can be done with equal ease and skill by males or females. The unit of study is one acre of pond i.e., labour employed to manage a shrimp pond of one acre.

Table 7.1 : Human Labour Employed in Shrimp Culture (Man days per acre-per Two Crops)

Farm Size	*Human Labour*		
	Male	*Female*	*Total*
Marginal	120.47	38.83	159.30
	(75.62)	(24.37)	(100.00)
Small	159.61	30.29	189.90
	(84.04)	(15.96)	(100.00)
Medium	176.00	22.00	198.00
	(88.89)	(11.11)	(100.00)
Big	201.62	24.78	226.40
	(89.05)	(10.45)	(100.00)
All Farms	164.42	28.97	193.39
	(85.01)	(14.99)	(100.00)

Note : Figures in paranthesis indicate the percentage to the total.

Human Labour Employment in Brackishwater Shrimp Culture

Labour is a primary factor of production. It is considered important not only because it is productive but also because it activates other factors and makes them useful for production purposes. Therefore, labour is or crucial importance in any economic activity.

The labour required to maintain shrimp farm in the study area is drawn from both family and hired sources. Hired labour is employed both on temporary and permanent basis. No child labour is directly engaged. The extensive type of culture system which is somewhat labour intensive is in operation in the entire study area. The family labour is monly coutribated by mates. Based on the data obtained from the sample farmers, the total employment generated per unit *i.e.*, per acre, in man days has been worked out and presented in Table 7.1.

The table shows, that there exists a positive relationship between the labour utilisation per unit and size of the farm. AS the size of the farm increases the labour utilisation increases. On an average, the employment generated in the sample shrimp farm is about 193.40 man days per unit area (one acre). Of this , male labour accounts for 85.01 per cent and the remaining 14.99 per cent is done by female labour. In respect of all the size groups the proportion of male labour to total labour employed is very high and is about 85 per cent. In the individual farms, the male labour is 75.62 per cent, 84.04 per cent, 88.89 per cent and 89.05 per cent respectively. It could be observed that the proportion of labour establishes a positive relationship with farm size. Thus the aquaculture of brackishwater shrimp appears to be a male dominated occupation.

Sources of Labour

Labour employed on the shrimp farms is drawn from family and hired (temporary and permanent) sources. The family labour component constitutes a substantially large proportion *i.e.*, about 70 per cent in small and marginal size groups due to the simple reason that these classes of farmers mostly consist of traditional fishermen communities and other weaker sections. These farmers would be able to undertake all farm activities with family members and presumably find no need to hire labour from outside. The size of the farms being small, they never call for labour use in excess of what the family members can provide at any point of time. In fact, there were instances where the labour available in the households far exceeded the requirement on these farms which might therefore result in excessive labour use on the farm. Perhaps this explains the vast difference in labour use that is observed between marginal, small farms and medium and big farms.

To ascertain the relative importance of family labour among different size classes of farms, the relevant data are presented in Table 7.2. In all the sample farms put together the family labour constitutes a higher proportion to the total labour used. The per cent of family labour to total labour is 53.38 per cent and the remaining 46.62 per cent is contributed by hired labour. But the family labour to the total labour in the case of medium and big farms is less *i.e.*, 48.98 per cent and 29.75 per cent respectively. When compared to marginal and small farms, this is significantly less. In the case of marginal and small farms the family labour is 69.17 per cent and 72.89 per cent respectively.

Another interesting point which the data reveal is that the proportion of hired labour to total labour increased in marginal, medium and big farms, but not in small farm. This is due to the fact that the sample farmers under this category belong to the traditional fishermen communities and therefore the involvement of their family members in the form of family labour is more in this class. As such the proportion of hired labour in this class of farm is less than what is in the other three categories, *i.e.*, marginal, medium and big farms. The proportion of hired labour in these three size groups to the total labour increases, establishing thereby a direct relationship between farm size and proportion of hired labour used. The percentage of hired labour in these three categories to the total labour is 30.83, 51.02 and 70.25 respectively.

The breakup of hired labour into male and female labour shows that the farmers expectedly are employing males when it is necessary to hire labour to supplement their family labour. Thus, on the whole it is noticeable from the data presented in the Table 7.2, that except in the case of small farm the proportion of family labour is decreasing as the size of the farm increases.

Table 7.2: Human Labour Employed Source-wise and Sex-wiss (Mandays Per Acre)

Farm	Family Labour			Hired Labour									Total Labour		
				Temporary Labour			Permanent Labour			Total Hired Labour					
Size	Male	Female	Total	Male	Female	Total	Male	Female	Total	Male	Female	Total	Male	Female	Total
Marginal	83.37	26.83	110.20	17.15	2.00	19.15	19.95	10.00	29.95	37.10	12.00	49.10	120.47	38.83	159.30
			(69.17)									(30.83)			(100.00)
Small	100.00	38.43	138.43	17.89	2.47	20.39	19.00	12.08	31.08	36.92	14.55	51.47	136.89	52.98	189.90
			(72.89)									(27.11)			(100.00)
Medium	92.00	5.00	97.00	25.11	4.36	29.47	57.56	14.00	71.56	86.67	18.36	101.00	174.67	23.36	198.00
			(48.98)									(51.02)			(100.00)
Big	64.92	2.42	67.37	46.13	6.25	52.38	91.03	25.62	116.65	127.16	31.87	159.03	202.08	24.30	226.40
			(29.75)									(70.25)			(100.00)
All Farms	86.74	16.50	103.24	26.51	3.77	30.34	45.63	15.42	61.05	71.96	19.19	91.15	158.52	34.86	193.39
			(53.04)									(46.83)			(100.00)

Note : Figures in Paranthesis Indicate Percentage to the total labour.

Female Labour Participation

In recent years, much attention was being paid to promote the role of women in the process of economic development by encouraging their work participation in India. Any attempt to alleviate poverty and increasing income and employment need to be based on an accurate understanding of the work patterns of men and women. Generally female work participation depends upon the cultural traditions, customs and patterns of social hierarchy. Economic status of course is the most important intervening variable in female work participation in the rural setting.

Table 7.3 : Percentage of Female Labour to Total Labour

(Mandays per Acre)

Farm Size	*Female Labour*	*Total Number*	*Percentage to total labour*
Marginal	38.83	159.30	24.37
Small	52.98	189.90	28.03
Medium	23.36	198.00	11.79
Big	34.30	226.40	15.15
All Farms	37.36	193.40	19.31

In this regard, we can find out about the female work participation in shrimp farming in the study area in Table 7.3. As can be visualised from the data, the rate of female participation is phenomenally low. For the sample as a whole the percentage of female labour to total labour force is 19.31 only.

As per the above table in the case of small farm group, which is dominated by fishermen communities the percentage of female labour is highest *i.e.*, 28.03. But in the case of marginal, medium and big farm groups the percentage of female labour to the total labour force is 24.37, 11.79 and 15.15 respectively. No significant relation is observed between the size of the farm and the participation of female labour.

As per the table the proportion of female labour has an inverse relationship with size of the farm. In big farm size group the female labour participation is increasing because both the wife and husband work as permanent labour. But the female labour participation in medium and big farm group is less when compared to small and marginal farms. Economic status thus seem to be an intervening variable affecting a decline in female work participation rate. However, from the increase in the percentage of female labour from big farm group, we can infer that profitable nature of shrimp farming is making the women to overcome the old traditions, customs and other barriers to participate in this activity.

SECTION-II

Marketing of Brackishwater Shrimp

Introduction

The ultimate aim of the producer in any enterprise is to get maximum return from the produce through efficient production and marketing, and the producers of shrimp are no exception. Like in all other enterprises the success and the economic viability of shrimp culture also depends upon its marketing system and practices.

In the case of inland fish culture, it is observed that there are many intermediaries who are engaged in the marketing process. Unlike cultured fish, cultured shrimp has been a major item in the seafood exports of India. It is hundred per cent export oriented item among the cultured species. India is recognised as one of the world leaders in the brackishwater shrimp culture production and export. There is persistent demand and ready market for the farmed shrimp of India in the global market. It is also observed that the shrimp market is generally simple and free from many intermediaries. The marketing costs are also said to be negligible i.e., about Rs. 5 to 10 per one kilogram of shrimp. However keeping in view the fraudulent practices in inland fish market an attempt is made to study the marketing structure of shrimp in the study area with particular emphasis on the marketing process.

Marketing Structure and Process for Brackishwater Farmed Shrimp

Till very recently the shrimp (Prawn) has been captured in seas, rivers, lakes, etc. Most of the shrimp captured was generally sold on the seashore, river banks etc., and the rest transported to the nearest domestic market for sale. However, the rapidly growing demand for shrimp in the international market on one side and the more or less stagnant shrimp harvest from natural sources on the other side, led to the introduction of aquaculture of shrimp in the brackishwater regions in India, during the last few years. Shrimp is generally considered as highly valued commodity and has been affordable to people in the higher income groups of developed countries like U.S., Japan and several other countries in Europe.

The domestic market of shrimp is negligible in India. The per capita consumption of shrimp has remained relatively low (less than 150 grams) over the past two decades. This is partly due to the fact that a large number of Indians are vegetarians and more importantly due to the high price of shrimp in relations to the purchasing power of a large majority of the population. In addition, over the last two decades, currencies of the major market countries have generally grown stronger in relations to those of shrimp producing countries in the developing world. Consequently, shrimp prices in many of the developing countries increased sharply in local currencies while real prices in developed countries that import shrimp from developing countries remained stable or declined during the same period.

This shift has meant that on an average shrimp consumption per capita has declined in developing countries while average per capita consumption in developed countries has increased. The global market for Indian brackishwater cultured shrimp has been on the upswing in terms of both volume and unit value and so the shrimp farmers of India are unlikely to face any marketing problems in the near future.

Harvesting of cultured shrimp is generally done when the shrimp reaches marketable size *i.e.,* after 3 or 4 months of rearing, preferably between the 4th and 8th day after full and new moon. Normally shrimp lies burried partly or completely at the bottom of the pond during the day time and become active during night hours by moving around in the pond for food. So the farmers generally harvest the shrimp during night hours. The normal size of the harvest in India is 20 grams for White Prawn (*P. indicus*) and 30 grams for Tiger Prawn (P. monodon). The harvested shrimp is sold in the headless form to the local purchasers who are the agents of processors or exporters located at Visakhapatnam or Madras. The harvested shrimp are collected by these agents at the farm site and they make their own arrangement for the transportation of shrimp to the nearest road-head from where the material is transported by insulated vans. Packing at the farm level is done in ice baskets made out of split-bamboo reapers. The shrimp thus packed reaches the processing centres where it is processed and exported.

The processing of shrimp is a standard procedure. On arrival from the producing centres, the shrimp is cleaned, graded and weighed in headless, shellon form in the sets of 1.8 Kg. or 2 Kg. The sets contained in wooden frames are placed in a freezer at 40°C (degree Celsius). In one hour and a half to three hours, depending upon the type of freezer, they turn into ice slabs weighing between 2.2 Kgs. and 2.4 Kgs. The slabs are placed in waxed duplex cartons with a polythene film lining inside. Each carton can accommodate about 6 slabs. Then, until use, the cartons are placed in cold storage where temperature is maintained at 20 degree Celsius. If properly maintained, the shrimp would lost at least 18 months without losing food value, quality or taste.

The Indian exports command a higher share of the world market mostly by volume, rather than by value, reflecting the lower price that Indian shrimp receive in the world market. This is due to the market perception, that Indian shrimp is of lower quality than those exported from other countries because of poor handling of shrimp. The quality of shrimp should be reasonably good, fresh and safe for consumption in order to attract better price in the international market. As most of the future shrimp exports will be coming from cultured shrimp, the farmers and the processors should carefully monitor the quality, so that India can compete in the world market which is becoming more and more quality conscious and price competitive.

The present market trends reflect a rapidly growing demand for ready to serve and ready to cook items. The importance of conventional frozen and canned products is getting diminished day by day. The sophisticated health conscious consumers abroad as well as the urban consumers at home demand new types of value added, hygienically prepared nutritious and a attractively packed products. Therefore the emphasis in future should be on the Individual Quick Frozen (JQF) products which fetch better price than the conventional block frozen materials in the foreign markets.

The Ministry of Commence, Government of India has identified marine products as a thrust area for export promotion. Marine

Export Development Authority was established by this Ministry to promote and diversify the marine products especially shrimp. As the market promotion is the motto of this organisation, MPEDA arranges seafood trade fairs and organises buyers and sellers meets. It also arranges programmes of export traders and provides opportunity for trading partners to meet, interact and strengthen their trade links.

To promote overseas trade in shrimp and other marine products MPEDA has opened offices at Tokyo, New York and Brussels. The Indian trade centre under the Indian Mission to the European Economic Community looks after and promotes seafood trade interests. It also draws expertise on international co-operation with international bodies like Food and Agriculture Organisation (FAO), European Economic Community (EEC), United Nations Council for Trade and Development (UNCTAD) etc., for providing services and finance to boost up seafood trade. Brook Bond, Voltas, Union Carbide, Indian Tobacco Company, Kings International and George Major are some important companies in the export trade of shrimp. In Andhra Pradesh companies like Indus Seafoods, VBC, Nikkanti are engaged in the export trade of shrimp. The brackishwater cultured shrimp from the State of Andhra Pradesh is exported mainly through Madras and Vizag ports.

Marketing of Shrimp

Two simple marketing channels are found in brackishwater shrimp marketing system. Channel (i) is producer-local buyer-cum-commission agent-cum-money lender-processor-exporter, Channel (ii) is producer-local buyer-cum-commission agent-processor-exporter. Nearly 53.03 percent of farmers sell their shrimp through the first channel. The remaining 46.97 per cent of them sell their produce through the second channel.

As for different size groups, the majority of the marginal farmers i.e., 64.28 per cent, sell their produce to local buyer-cum-commission agent—money-lender as he provides credit to them. Therefore, the farmers are compelled to sell their produce to him at the farm level 51.21 per cent of small farmers also sell their produce through the first channel. Only 30 per cent of medium farmers and 20 per cent of big farmers sell their produce through the first channel. These farmers sell their produce through the

Table 7.4 : Details of Shrimp Sold by Size Group and Channel-wise.

Farm Size	Local agent-cum-buyer	Quantity sold in Kgs.	Local agent-cum-buyer-cum-lender	Quantity sold in Kgs.	Total	Total quantity sold in Kgs.
Marginal	35	7,525	63	15,948	98	23,472
	(35.72)	(32.05)	(64.28)	(67.85)	(100.00)	
Small	40	19,003	42	26,483	82	45,486
	(48.78)	(41.78)	(51.21)	(58.22)	(100.00)	
Medium	14	20,113	6	7,057	20	27,170
	(70.00)	(74.02)	(30.00)	(25.98)	(100.00)	
Big	12.00	45,205	3	9,735	15	54,940
	(80.00)	(82.29)	(20.00)	(17.71)	(100.00)	
All Farms	101	68,004	114	2,10,064	215	1,51,068
	(46.97)	(45.01)	(53.03)	(55.09)	(100.00)	

Note : Figures in the paranthesis indicate per centage to the total.

second channel in which they do not need to sell their produce on compulsion to the agents as they do not generally borrow any money from the buyer-cum—agent. The farmers pay a commission somewhere between Rs. 3 to 5 per Kg. of shrimp to the agent.

Marketing Problems of Cultured Shrimp Farmers

The following problems are identified in the marketing of farmed shrimp.

1. The farmers are placed at a disadvantage in regard to marketing of cultured shrimp. The farmers feel that the prices in the adjoining Tamil Nadu area are far more attractive and reasonable than in Andhra Pradesh. Moreover the system of their fixing the prices treating any fraction within 10 per cent as 30 count is much more remunerative than the prevailing system in Andhra Pradesh. In the study area, separate prices are fixed for 20 count, 21 count,22 count and so on. In other words, shrimp with 21-25 count fetches a price equal to 20 count while 26-29 count is not equated with 30 count but has a price of its own. The shrimp bring returns to the farmers according to the count level. If smaller shrimp constitute a major part of the harvested shrimp, the average returns go down. Due to the prevailing system of price fixation of shrimp, the farmers of Andhra Pradesh in general and the farmers

of the study area in particular are deprived of getting a reasonable price for their cultured shrimp.

2. The farmers sell their shrimp to the purchasing agents of exporters, located at Vishakhapatnam or Madras. Agents collect the harvested shrimp at the farm site and they make their own arrangement to transport the shrimp to the nearest road-head from where the material is transported in insulated vans. Though there is no problem of incurring marketing costs directly, the agent collects a flat rate of Rs. 5 for each Kg. of shrimp sold by the farmer towards marketing expenses. In addition, the agent collects Rs. 5 on each Kg. of shrimp purchased by him from the farmers as his commission.The farmers have expressed the feeding, that these agents are middlemen who represènt the processors or exporters invariably offer low rates and exploit them. This would also help the local agents to get better commission from the exporters.

3. Another practice observed in shrimp marketing in the study area is that the local purchaser or agent takes advance from the exporter one or two months before the harvest and lends it to the farmers at an interest rate of Rs. 3 per hundred per month. He also makes it obligatory to all the farmers who borrow money from him to sell their produce through him. So that he can deduct his commission and the credit given with interest from the sale proceeds before the final payment is made to the farmers.

4. The road and communication facilities are not good in the study area, especially in Kona village of Bandar mandal and Tadivennu village of Kruthivennu mandal. In summer there would not be much difficulty, but during rainy season to cross the creeks, cannals and rain affected roads becomes extremely difficult because of the absence of proper roads and bridges. This ultimately increases the cost of transportation of building material, feed and other equipment required to the farm site.

Conclusion

The pattern of employment in brackishwater shrimp culture largely depends upon the type of culture practices. The semi-intensive, and intensive type of culture are more techno-capital intensive than labour intensive. But under the Extensive type, the farmers use more labour and relatively less machinery as the stocking density of seed is less under this type of culture. In the study area only extensive type of culture is found to be in operation. On average the employment generated in shrimp farming is about 193 mandays per acre from two crops in a production year. It is observed that as the size of the farm increases labour utilisation increases. Of the total labour, male labour accounts for more than 80 per cent. Family labour component is about 70 per cent, while the hired labour constitutes the remaining 30 per cent. The family labour constitutes high per cent in the case of small and marginal farmers as most of them belong to fishermen communities and other weaker sections.

Shrimp marketing involves less number of middlemen when compared to fish marketing. The rate of exploitation is less. The marketing expenses of the producer are negligible. However, the farmers are not happy with the system of price fixation, and available infrastructure.

Chapter 8

Summary and Conclusions

The importance of fishery sector in developing countries is now well-known. In fact, it has a place of pride in the national economy of India. But as the harvest of fisheries from natural sources remained more or less stagnant, the accent rightly fell on aquaculture of fisheries. Aquaculture of fish or fish farming therefore emerged as an alternative method of augmenting fish production in the country. Aquaculture has also been recognised as one of the best allied activities of agriculture from the stand point of income and employment generation to the rural poor

'Shrimp' which is an aquatic organism inhabiting the sea, estuaries and backwaters remains the backbone of Indian marine products export. Shrimp is a protein-rich food with universal taste. Shrimp farming is a short duration activity and therefore provides quick return on investment. It also provides gainful utilisation of derelict and unproductive coast swamps and generate employment opportunities to the rural poor in remote coastal areas. The value of shrimp exported in 1993-94 from India was estimated at about Rs. 2,150 crores (U.S.$ 650 million) out of the country's total export of Rs. 69,571 crores (U.S $ 25,541 million). It is estimated that one hectare of shrimp culture operation can provide employment to at least two persons. It can also provide indirect employment to the extent of 1100 mandays. If one lakh hectares of shrimp culture is developed, it generates direct employment for 2 lakh persons and indirect employment to another 3-5 lakhs according to MPEDA.

Although fishing is an age-old, caste-based activity in India, aquaculture of fisheries is of recent origin. The aquaculture of brackishwater shrimp has emerged as commercial and scientific activity in India only in the early eighties. Until recently shrimp production has been exclusively from marine capture and traditional brackishwater areas like 'pokkali' fields and 'bheries'. But India is now recognised as one of the world leaders in shrimp farming and the Government of India has identified shrimp farming as 'thrust area' for export promotion because of its persistent demand in the world market. Therefore in India commercial aquaculture has been considered synonymous with brackishwater shrimp culture.

India is endowed with rich natural resources for shrimp farming in coastal zones in the form of brackishwaters, swamps, lagoons, mangrove backwaters and estuaries. The area suitable for shrimp culture is estimated at about 1.2 million hectares of which 82,000 hectares has already been brought under shrimp culture.

Shrimp farming has been classified into four types on the basis of stocking density of seed and method of cultivation. They are : (1) traditional farming : (2) extensive farming; (3) semi-intensive farming; (4) intensive farming. In India the first two types and to some extent the third type of culture system are in practice. Semi-intensive type is confined only to big farms and corporate companies. The extensive system is in extensive use among small and marginal farmers. In India more area under brackishwater shrimp culture is under the extensive type. In the entire study area only extensive type of culture is found to be in practice. As the water salinity and environment in the West Coast is not very suitable, the aquaculture activity of brackishwater shrimp is mainly confined to the East and South Coasts comprising the States of West Bengal, Orissa, Andhra Pradesh, Tamilnadu, Karnataka and Kerala.

Owing to the abundance of natural resources, the country offers immense investment opportunities in shrimp farming. Impressed by the scintillating performance of the shrimp in the global market, the Government of India allowed the Ministry of Commerce to formulate several schemes intended to provide infrastructural, financial and technical facilities to the shrimp farmers so as to induce them further into this new and profitable venture. Shrimp farming in India has taken new dimension in recent years largely due to the promotional programmes launched by the MPEDA.

Besides technical assistance, it also provides subsidy for shrimp culture and hatchery development. Under the management of

MPEDA to commercial shrimp hatcheries, one at Gopalpur i.e. Orissa Shrimp Seed Production and Research Centre (OSSPARC) and the other at Visakhapatnam *i.e.*. The Andhra Pradesh Shrimp Seed Production and Research Centre (TAPSSPARC) with overseas technical-know-how have been established. Besides, Brackishwater Fish Farmers Development Agencies (BFFDA) are also extending their support to small shrimp farmers by way of technical input and subsidy. In addition, several nationalised commercial banks, co-operative societies are providing credit which is normally refinanced by National Bank for Agriculture and Rural Development (NABARD) to promote shrimp farming. The financial institutions like Shipping Credit and Investment Corporation of India (SCICI) and Industrial Credit and Investment Corporation of India (ICICI) are providing credit facility to the integrated, export oriented shrimp farming units in the country. Various Universities, Fisheries Colleges and the Indian Council of Agricultural Research (ICAR) have been actively involved in developing appropriate technologies for shrimp farming.

In view of the vast potential offered by this sector and the deligent and unstinted organisational support given by the government, the pace of development of brackishwater shrimp culture has been impecable and fairly rapid over the past few years in India. Some of India's large private companies have already invested sizeable amounts of money in setting up shrimp farming projects. The concept of 'Satelite Farming' is also coming up in all maritime states. It would appear from the foregoing account that India possesses a rich potential for shrimp farming and that the efforts are being made to exploit the resources to the optimum level which may bring a remarkable break through in shrimp production and export in the near future. 'Aquaculture Foundation of India' which has been established recently is playing a significant role in this aspect.

The studies (Vide Chapter-4) available in the existing literature on this subject are very few and quite inadequate. However, these studies have raised certain pertinent issues. They are : (i) Aquaculture activities are more income yielding and profit generating than agriculture; (ii) technical and managerial problems are the main constraints of development of aquaculture; (iii) there are several marketing problems of culture fisheries; (iv) there are differences in net income in capture and culture fisheries; (v) there are several factors relating to costs and returns, input-output relationships. But none of these studies are indepth with empirical base on brackishwater shrimp culture. Hence the present study is designed to fill this gap. An attempt is therefore made to evaluate the costs and returns in brackishwater shrimp culture, input-output relationship, employment generation and market structure.

Of all the nine coastal districts in Andhra Pradesh, Krishna District is having more area under brackishwater shrimp culture. Therefore this distinct is selected purposively for the present study. Out of 50 Revenue Mandals in the District, seven Mandals are located along the coast line and out of these seven, shrimp culture is in operation in six Mandals. From out of these six Mandals three Mandals viz., Bandar, Bantumilli and Kruthivennu have been chosen for the study because more than 50 per cent of brackishwater area is within these three mandals.

From among these three mandals, one village from each mandal has been selected. These villages are Kona in Bandar Mandal, Manimeshwaram in Bantumilli Mandal and Tadivennu in Kruthivennu Mandal. The selection of the villages is based on the strength of the farmers who have experience in shrimp farming for a considerable period of time and their representative nature from various size groups. The three selected villages together have 430 shrimp farmers belonging to various size groups. The farmers are classified into four categories namely marginal (below 2 acres), small (2-5 acres), medium (5-10 acres) and big (above 10 acres) depending upon their land holding in shrimp culture. Out of these 430 farmers belonging to various size groups 215 farmers i.e. 50 per cent are selected for the study based on the probability, proportional random sampling technique i.e. in proportion to the number of farmers in total farmers of a size group. Finally, random technique is employed to select the sample farmers from each group, wherever the data is presented in tabular form the simple technique of tabular analysis is used and the results are presented in terms of averages and percentages. The other important statistical technique used is regression analysis. This technique is employed to analyse the factors affecting brackishwater cultured shrimp production. The whole thesis is divided into eight chapters.

Main Findings

The major findings of the study are divided into the following five parts :

1. Growth of brackishwater shrimp farming in Andhra Pradesh.

2. Factors inducing the farmers into shrimp farming, and the socio-economic profile of the sample farmers.

3. Employment potential in brackishwater shrimp culture.

4. Marketing structure of the brackishwater cultured shrimp.

5. Costs and returns in brackishwater shrimp culture.

Growth of Brackishwater Shrimp Fasming in Andhra Pradesh

(a) Fisheries Resources

The aquatic resources of Andhra Pradesh are varied and extensive and are broadly divided into two types : (1) Marine and (2) Inland. Andhra Pradesh is the second largest maritime state in the country, having a coast line of about 982 Kms. The estuaries, backwaters, bays and brackishwater lagoons form an important brackishwater culture sources of the state. The ecosystem in the state with transitory zones between freshwater and off-shore seawaters are highly productive to culture the valuable fisheries like mullet, hilsa and highly valuable shrimp.

(b) Availability of infrastructure facilities

In the State of Andhra Pradesh there are about 45 processing plants with a total capacity of 4900 MT of shrimp. There are about 135 registered export companies of shrimp. There are 27 hatcheries (25 under private management and 2 under public sector) in the State. A number of hatcheries are in the offing. Since the hatcheries are unable to supply adequate quantities of seed, farmers continue to depend upon wild sources of seed to a large extent.

There are no complaints of shortage of feed but there are complaints about the quality and performance of imported feed. Majority of the farmers prefer indigenous feed to imported feed. Though no serious transport bottlenecks are reported in the study area it is observed that during the rainy season it is very difficult for the vehicles to cross the creeks as there are no bridges. Most of the farmers involved in this activity have no clear knowledge of new and scientific methods of farming. Therefore, the farmers are emphasising on the need to be provided with training and extension facilities. Except for a little delay in the payment of cash to the shrimp sold, no other serious problem of marketing is reported in the study area by the farmers. However, the farmers have some dissatisfaction about the system of price fixation. The farmers feel that the price fixation system of adjoining Tamilnadu State is far better when compared to Andhra Pradesh where the study area is located.

(c) Growth of brackishwater shrimp culture in Andhra Pradesh

Andhra Pradesh stood in the first place in cultured shrimp production with 26,000 tons (41.93 per cent) and in the second position in the area with 19,500 hectares (23.62 per cent) in the country.

To promote shrimp culture in Andhra Pradesh MPEDA has opened its Regional Offices at Visakhapatnam, Vijayawada and Bhimavaram. At Visaknapatnam TASSPARC which was established by MPEDA has been operating with a capacity of 100 million seed production per annum with overseas technical know-how. There are six BFFDA's established by the Central Government at the following coastal districts, *viz.*, Krishna, Nellore, Prakasam, West Godavari, East Godavari and Srikakulam. Out of the 31 feed producing units 13 are in Andhra Pradesh and most of them are in Vijayawada of Krishna District. Out of the 54 hatcheries in the country 27 are in Andhra Pradesh. Out of the 48 hatcheries under construction 35 are in Andhra Pradesh (MPEDA).

Of late, some service organisations like Faith Action for Community Transformation (FACT) and Churches Auxiliary for Social Action (CASA) have also involved in shrimp culture through rural poor with the assistance of District Rural Development Agency (DRDA). There are about 6 brackishwater farmers co-operatives in the district, formed to safe-guard and promote the interest of the shrimp farmers. The FAO and World Bank sponsored shrimp and fish culture project with an estimated cost of Rs. 83.13 crores is in offing in the state.

Factors Inducing the Farmers Into Shrimp Farming and the Socio-Economic Profile of the Sample Farmers

Literacy is one of the social indicators of development. It is generally believed that aquaculture of shrimp on scientific lines would require a certain level of education. Majority of the sample farmers are found to be literates *i.e.* about 56 per cent of them are literates. 40 per cent of the farmers have education upto primary level, followed by secondary level 9.30 per cent, graduate level 5.11 per cent and technical education for 1.4 per cent. The finding thus indicates that the majority of the shrimp farmers who have undertaken shrimp farming do have the necessary level of formal education.

The selected farmers took up shrimp farming both as principal and secondary occupation . About 80 per cent of the sample farmers undertook shrimp culture as their principal occupation and the remaining of them as their secondary occupation. It is contextual to note that the majority of the sample farmers with shrimp culture as their principal occupation are marginal and small farmers. This may imply that these farmers have taken up brackishwater shrimp culture, as most of them are allotted derelict coastal lands by the government and majority of them are fishermen by occupation.

Almost all the sample shrimp farmers with the exception of two in Tadivennu village of Kruthivennu Mandal are following mono-culture of Tiger shrimp. These two farmers are raring P. indicus (white prawn) to about 10 per cent along with Tiger shrimp on trial basis. All the sample farmers are owner farmers.

Since shrimp farming is comparatively a new venture, factors that are inducing the farmers to enter into this venture are observed. Based on the responses given by the sample farmers, four important factors are identified : (1) Availability and suitability of land; (2) profitability; (3) Seepage factor and (4) other factors. Availability of suitable land is found to be one of the compelling reasons for many (61.86 per cent) of the sample farmers to plunge into this activity. 20.93 per cent of the farmers took up this activity on profitability criteria and the ramaining 17.20 per cent due to seepage problem i.e. intrusion of salt wafter into adjacent paddy fields.

Since aquaculture of shrimp is a new venture in the State and also in the study area, emphasis has been placed on the experience of the farmers to collect the relevant data. The scholar felt that the farmers in the study area on average must have at least 3-4 years of experience in shrimp farming. After careful survey and a number of visits and interactions with farmers the study area is decided. The data finally show that more than 50 per cent of the sample farmers on average have more than 4 years of experience in brackishwater shrimp culture. This justifies the selection of the sample farmers on the basis of their experience for a period of at least 3-4 years.

The caste composition of sample shrimp farmers reveals an interesting fact that almost all caste groups have entered into brackishwater shrimp farming. Shrimp culture has broken all caste barriers and attracted and induced farmers from all castes because this enterprise is found to be more lucrative and remunerative than other agricultural activities.

Seed is an important input in shrimp culture. In the study area, two sources of seed are identified. They are : (1) Wild seed, (2) hatchery seed. On an average 75.36 per cent of sample farmers obtain seed from natural or wild source (i.e. from backwaters, estuaries, coastal waters), 6.97 per cent from hatcheries and 17.67 per cent from both the sources. The data reveal that the natural seed is still the chief source of seed pointing to the fact that we have still to traverse a long part to change this pattern.

Employment Potential in Brackishwater Shrimp Culture

Regarding labour absorption in shrimp farming, on an average, the total employment generated in the sample shrimp farms is about 194 mandays per acre for two crops. Of this, male labour accounts for as high as 85 per cent and the remaining 15 per cent for female labour. In respect of all the size groups, the proportion of male labour to total labour employed is very high and is about 85 per cent. Thus, the aquaculture of shrimp appears to be a male dominated occupation. Considering all the sizes together, we find that the proportion of hired labour is less than the family labour i.e., family labour is 53.38 per cent and hired labour 46.62 per cent.

Regarding the labour utilisation according to the size groups the per cent of family labour in big farms is 29.75 per cent and in medium farms 68.98 per cent. In the case of marginal and small farms the family the labour is 69.70 and 72.89 per cent respectively. Except in the case of small farms in all the other three size groups marginal, medium and big, a direct relationship is observed between farm size and hired labour. This is due to the fact that most of the small farmers belong to the fishermen community constitute 72.89 per cent of their family labour and only 27.11 per cent hired labour in shrimp culture.

Marketing Structure of the Brackishwater Cultured Shrimp

Two simple marketing channels are found in barckishwater shrimp marketing system in the study area. Nearly 53.03 per cent of farmers sell their shrimp through the first channel *i.e.*, producer-local buyer-cum-commission agent-cum-money lender-processor-exporter. The remaining 46.97 per cent of them sell their produce through the second channel i.e. producer-local buyer-cum-commission agent-processor-exporter.

As for different size groups, the majority of the marginal farmers i.e., 64.28 per cent, sell their produce to local buyer-cum-commission agent-cum-money lender as he supplies credit to them. Therefore, the farmers are compelled to sell their produce at the farm level. 51.21 per cent of small farmers also sell their produce through the first channel. Only 30 per cent of medium farmers and 20 per cent of big farmers sell their produce through first channel. These farmers generally sell their produce through the second channel in which they do not need to sell their produce on compulsion to the agents as they do not generally borrow any money from them.

Costs and Returns in Brackishwater Shrimp Culture

It is seen that on an average the total non-recurring capital per acre of shrimp pond worked out at Rs. 16,979 for the sample as a whole. The total non-recurring capital per acre varies from Rs. 12,989 to Rs. 22,988 from the marginal farmers to big farmers. It is further observed that the farm size and non-recurring capital per acre are directly related.

With regard to the recurring or working capital of production, for the sample as a whole the recurring capital of production per acre of shrimp pond was Rs. 29,545. Recurring capital per acre also exhibits direct positive relationship between the size of the farm and recurring capital among all farm size groups.

Among different recurring cost items, feed alone accounts for about 36 per cent. Marked relationship is observed between feed cost and farm size. The feed cost increases with the increase in the size of the farm. Next to feed, expenditure on seed constitutes the major share of about 29 per cent, followed by labour cost of 18 per cent, fuel cost of about 7 per cent, fertilizer cost of about 3 per cent, cost of pesticides and expenses on miscellaneous items varying from 3 per cent to 2 per cent.

The total cost of production (fixed and variable cost) for the sample as a whole is Rs. 33,770. The total cost increases with the increase in the size of the farm from Rs. 28,821 on marginal farms to Rs. 37,654 on big farms. Direct relationship is observed with regard to the production cost and the size of the farm. In the total production cost, variable costs constitutes 88.40 per cent while the fixed costs constitutes 11.60 per cent. Among the variable cost items the feed cost is the highest with 32 per cent, followed by seed 25.61 per cent, labour 16.50 per cent, followed by seed 25.61 per cent, labour 16.50 per cent, miscellaneous expenditure 7.97 and fuel 6.32 per cent respectively.

The net returns per acre, per year from (two crops) for sample as a whole is worked out at Rs. 50,008. The net returns increases with the increase in the size of the farm establishing a direct relationship between the net returns and the size of the farm. The output-input ratio between various size groups vary from 2.52 per cent on marginal farms, 1.44 per cent on small farms, 1.55 per cent on medium farms and 1.50 on big farms with an overall average of 1.48 per cent. The input cost in relation to output is less among marginal farmers. The marginal farmers are able to get more yield with less input cost. However, the input-output ratio in the other size groups is slightly varied. This is because of the scientific nature of shrimp culture where the farmers irrespective of their size groups use the same inputs in almost the same quantity.

The production function analysis reveals that in brackishwater shrimp culture supplementary feed, seed and miscellaneous items are the important factors that affect the gross returns. In this context, multi-collinearity tests are observed and it is found that there is no linear relationship among the explanatory variables.

Policy Implications

Brackishwater shrimp culture has certainly emerged as 'Sun Rise' industry (Shaktivel) to tap enormous resources of bio-energy available in the country for the benefit of rural poor especially in the coastal zones. This fascinating field of shrimp culture would set up strong base for opening new frontiers in health food production, employment and income generation for improving the quality of life of rural people particularly small and marginal farmers and landless labour. The findings on cost and returns and employment generation supports this point. In view of the vast potential offered by this sector, the pace of development has also been fairly rapid over the last few years in India. In the process, some undesirable effects are bound to occur. Therefore, there is urgent need for comprehensive and cautious policy approach for the healthy and eco-friendly expansion of this activity. Based on the findings of the present study, a few measures which may help in designing the programmes and formulating policies relating to brackishwater shrimp culture development are suggested below. However, it is necessary to point out that the results of this study should be taken with necessary caution since the sample is quite small and represents only a particular area.

(1) Since the brackishwater shrimp culture is a new and scientific venture, there is an urgent need to provide necessary training to the farmers. In the absence of any such facilities provided by the government, farmers are depending upon pseudo-consultants who offer wrong advice to farmers. The agencies like Marine Products Export Development Authority, Brackishwater Fish Farmers Development Agencies and the Department of Fisheries therefore should initiate programmes to provide necessary training extension, infrasturcture and technical guidance to the farmers on the use of feed, seed and water management at least at crucial points as the need for such help has been repeatedly voiced by the farmers.

(2) Brackishwater shrimp culture has to be developed as eco-friendly activity as there are complaints regarding pollution of water and environment. In this regard, the State Government should insist on obtaining clearance certificate of State Pollution Control Board (SPCB) and incorporate environment safeguards for warding off adverse impact of shrimp culture on environment.

(3) Indiscriminate conversion of agricultural lands and mangroves into shrimp farming in coastal zones is reported to have been causing imbalance in land use effecting the eco-system and salt intrusion into drinking water wells and making the water unsuitable for drinking. In this regard, the government has to take up necessary legal measures to regulate the unplanned and haphazard expansion of shrimp culture.

(4) Availability of natural seed which is highly seasonal is the most important input in shrimp culture. The seed produced and supplied through hatcheries is rather limited. Therefore, the farmers often fail to procure enough quantity of seed during each season. To combat this problem new hatcheries have to be established with required level of efficiency both in the public and private sectors.

(5) The export performance and value realisation would depend upon value addition and quality of shrimp products. Therefore, the government has to perforce to introduce quality standards in all aspects of aquaculture operations of shrimp right from site selection to the product packaging and make sure that World Trade Organisation (WTO) related non-tariff barriers, such as environmental issues, if and when applied, do not effect our shrimp culture production and export. These measures would help the farmers and export traders to get much better price for the cultured shrimp.

(6) Recently a wide spread attack of diseases on Tiger shrimp under culture has been reported in most of the farms in Andhra Pradesh including the area under study. Though the situation has not reached a point of no return to the farmers as yet, the farmers and the Government cannot afford to be complacent and continue to ignore the problems. In this context it is to be noted that the concentration on 'Tiger shrimp' culture alone is not desirable and there should be a diversification in culture practices. Adequate importance should be given for culture of broad spectrum of shell-fish and fin-fish like, crabs, clams, mussels, lobster, pearl culture etc., instead of restricting it to only one or two species like Tiger prawn and White prawn as at present.

(7) Since the shrimp farming is also a high risk area, the Insurance companies should revise the insurance schemes to provide insurance cover both for total and partial losses in aquaculture of shrimp. The recent viral attack on shrimp under culture, the consequent loss sustained by the farmers emphasises this point. As per the reports available, the insurance industry has already developed a scheme for shrimp farming. There is an urgent need to implement such schemes in brackishwater shrimp farming. All this required a comprehensive package of programmes by the government embracing all the above considerations. It should carry out all these programmes simultaneiously in order to achieve better results from brackishwater shrimp farming in the country.

Appendixes

Appendix I
Is It 'Prawn' Or 'Shrimp'

The standardisation of the term shrimp and prawn was the subject of discussion at the 'Prawn Symposium' of Indo-Pacific Fisheries Council held at Tokyo in 1955. In this meeting the IPFC resolved that the word 'prawn' should be applied to Penaeids, Pandalids and Palemonids and 'shrimp' to the smaller species belonging to the other families (Kurian and Sebastian, 1982, p. 75).

At the 1967 World Conference on the 'Biology and Culture of Shrimp and Prawn' held in Mexico City, it was agreed that the term prawn be used for freshwater species only and their marine and brackishwater relatives be termed as shrimp (Bay of Bengal News Information Letter No. 33016/5/91 FY (2), Government of India, Department of Agriculture and Cooperation, Krishi Bhavan, New Delhi received by Marine Export Development Authority).

As per Aquaculture Extension Manual 5 (ISSN-0115-5369, Manual on Prawn Farming, July 1979, South East Asian Fisheries Development Centre, (Philippines), there is actually no clear distinction between the terms shrimp and prawn, except according to the dictionary definition. As per the Dictionary definition, shrimp is the small side of prawn while prawn can range from one inch to the size of a lobstar.

Marine Products Export Development Authority in its magazine 'Aquaculture Drops', Vol. III, Issue No. 1, July 1990, gave the following clarification on the question 'What is shrimp and what is prawn'. Quoting the clarification given by Food and Agriculture Orgnisation, MPEDA says, shrimp is the species raised in sea and brackishwater and prawn refers to the species raised under freshwater conditions. But in Indian context shrimp and prawn are used interchangeably.

Appendix II
Shrimp Exports from India

(1981-82 to 1992-93)

Year	*Quantity in tonnes*	*Value in Rupees (Crores)*
1981-82	52,180	247.95
1982-83	55,002	316.15
1983-84	54,444	314.81
1984-85	55,398	329.70
1985-86	50,349	329.82
1986-87	49,203	377.93
1987-88	55,736	425.78
1988-89	56,835	470.33
1989-90	57,839	463.36
1990-91	62,395	666.33
1991-92	73,107	976.16
1992-93	74,393	1,180.26

Source : (1) Hand Book on Shrimp Farming by MPEDA, Kochi, 1992.
(2) Royyalasagu-Melakuvalu, MPEDA, KOCHI, 1993.

Note : The shrimp exports from India between 1981-82 and 1992-93 have increased at a compound annual growth rate of 2.87 per cent and 13.19 per cent respectively.

Appendix III

Area and Production of Shrimp in Andhra Pradesh

Year	*Area brought under culture in hectares*	*Estimated production in M. Tonees*
1982-83	64.40	12.00
1983-84	119.20	30.00
1984-85	196.30	70.00
1985-86	370.18	140.00
1986-87	775.18	350.00
1987-88	1,384.32	940.00
1988-89	1,285.47	1,500.00
1989-90	2,691.47	2,850.00
1990-91	6,000.00	7,350.00
1991-92	8,100.00	9,700.00
1992-93	9,500.00	12,800.00
1993-94	19,900.00	26,000.00

Source : MPEDA Regional Office, Vizag.

Note : The area brought under shrimp culture and production of shrimp in Andhra Pradesh have increased between 1982-83 and 1993-94 at a compound annual growth rate of 67 and 101.28 per cent respectively.

Types of Shrimp Farming

According to the nature of scientific management and inputs, shrimp farming systems can be broadly classified as traditional, extensive semi-intensive and intensive. However, there is no universally accepted classification for types of shrimp farming.

Traditional

In the traditional system of shrimp farming, shrimp seeds along with fish seeds etc. are trapped during hightide when the ponds get inundated. In this process many predators also get free entry and they all co-exist in the same pond. Escape of these organisms is prevented by fixing suitable screen (s) in the sluice and the crop is harvested at frequent intervals. Shrimp is often harvested before they grow to optimum size. Owing to indiscriminate stocking of both desirable and undesirable varieties of shrimps, fishes etc. the production under this system is normally unpredictable and often very low in quantity because of predation by fishes etc., low stocking density and frequent harvest. The average production from traditional system is below 0.5 tonnes/ha/annum.

Extensive

The extensive system of shrimp farming is an improved method of traditional farming, involving construction of new ponds ranging from 1 to 5 ha. in suitable selected areas, selective stocking with fast growing shrimp seeds at a comparatively lower density ranging from a few thousands to 1,00,000 seeds per ha. with supplementary feeding. The water quality is maintained either through the natural fall and rise of tides or exchange through pumping on a low scale (upto 10 per cent). The average production under this system normally ranges from 1 to 1.5 tonnes/ha/crop.

Semi-Intensive

The semi-intensive system of shrimp farming involves construction of ponds ranging form 0.2 to o.5 ha. in size, selective stocking with fast growing hatchery seeds at a comparatively high density ranging from 1 to 3 lakh per ha. maintenance of water quality by exchanging 10-20 per cent daily, aerating of the pond with air blowers/paddle wheels and feeding the shrimps with nutritious feed. The average production in this system ranges from 4 to 5 tonnes/ha/crop in 4.5 months.

Intensive

The intensive system of shrimp farming involves construction of concrete ponds of 0.03 to 0.1 ha. in size, selective stocking with quality shrimp seeds procured from hatcheries at a density ranging from 5 to 10 lakhs per ha. maintaining water quality by exchanging over 30 per cent a day, aerating the pond with mechanical aerators and feeding the shrimps with nutritionally well balanced high energy feed. The production from this system ranges from 10 to 20 tonnes/ha/crop.

The extensive system of shrimp farming is widely practised in the majority of the shrimp farming countries. But in recent years, the semi-intensive system of shrimp farming, especially those using pure seawater is making rapid strides, intensive farming is found to be highly risky due to disease and pollution problems and therefore this system is not popular despite high returns. Since there is no treatment of disease that affect shrimps and the area of intensive farming is likely to get polluted due to discharge of heavy load of metabolites it is advisable to go in for semi-intensive shrimp farming with a stocking density of about 2 lakhs/ha.

However in this study area only the Extensive type of shrimp culture with an average stocking density of 5000 to 6000 of seed is in operation.

Bibliography

Allen, R.G., et al. 1987. : *Bio-Economics of Aquaculture*, Vol. 13, Elseviour Science Publishers, Amesterdam, Netherlands,

Bal, D.V. and Rao,1984 : *Marine Fisheries*, Tata Macgraw Hill Publishing Company, New Delhi.

Blanco, C.J. and Pillary;T.V.R. 1972 : *Economic Trends of Coastal Aquaculture in Philippines*, Coastal Aquaculture in Indo-Pacific Region, London, Fishing News Books Limited.

Bose, A.N., et al.1991 : *Coastal Aquaculture Engineering*, Oxford and IBH Private Limited, New Delhi.

Chopra, B.N. (ed.) 1951. : *Hand Book of Indian Fisheries*, Agriculture Series No. 44, Ministry of Agriculture, New Delhi

Commissioner of Fisheries 1993. : *Handbook on Fisheries Stratistics of Andhra Pradesh*, 1992 93, Government of Andhra Pradesh, Hyderabad.

David, R.H. and Christopher T.H. 1989. : *Inland Aquaculture Development Handbook*, Fishing News (Books) Osney Mead Oxford OXZOEL, Britain,

Emmanuel D'Silva (ed.) 1982. : *Fishery Sector Policy*, The World Bank, Washington, U.S.A.,

Evan Brown, E. : World Fish Farming, Cultivation and Economics, The Avi Publishing Company Inc., West port, 1977.

Food and Agriculture Organisation,1966. : *Fish Culture in Central East Africa*, Rome,

F.A.O. : Fisheries in Food Economy, Basic Study, Note 19, Rome, 1968.

F.A.O. : Year Book of Fishing Statistics, Vol. 63, Rome, 1986.

F.A.O. : A Quarterly update based Market Trends Globofish Data Bank, Globofish 4/94, Rome, 1994.

Frederick, E. Croxton, et al. : *Applied General Statistics*, Prentice Hall of India, New Delhi, 1973.

Heady Earl, O., and Dillan, J.L. 1964. : *Agricultural Production Functions*, Iowa State University Press, Iowa.

Jhingran, V.G. : *Fish and Fisheries of India*, Hindustan Publishing Corporation, New Delhi, 1983.

Klein, L.R. 1973. : *Introduction of Economatrics*, Prentice Hall of India, New Delhi.

Koutsoyiannis, A. 1979. : *Theory of Econommatrics*, Macmilan Press Ltd., London

Kurian, C.V. and Sebastian, V.O. 1982. : Prawns and Prawn Fisheries of India, Hindustan Publishing Corporation, New Delhi.

Leleuma : The Design of Rural Development Lessons from America, John Hopkins, Baltimore, 1975.

Marine Products Export Development Authority 1992. : *Hand Book on shrimp Farming*, Kochi, Kerala, India.

M.P.E.D.A. 1993. : Royyalasagu-Melakuvalu Kochi.

M.P.E.D.A. 1993. : *Hand Book on Aquafarming*, Kochi.

Milne, P.H. 1972. : Fish and Shellfish Farming in coastal Waters, Fishing News Book Ltd., England,

Misra, G.P. 1982. : Dynamics of Rural Development in Village India, Ashish Publishing House, New Delhi.

Pillay, T.V.R. , 1977. : *Planning of Aquaculture Development-An Introductory Guide*, Fishing News Book Ltd., England.

Prasad, K.N. 1991. : *India's Rural Problems*, Concept Publishing Company, New Delhi.

Rao, P. S. , 1983 : *Fishery Economics and Management in India*, Pioneer Publishers, Bombay.

Santhanan, R., et al. 1990. : *A Manual of Freshwater Aquaculture*, Oxford and IBH Publishing Company Pvt. Ltd., New Delhi.

Shao-Wen Ling 1977. : Aquaculture in South East Asia : A Historical Over View, A Wahington Seagrant Publication, USA.

Shigueno, K. 1975. : Shrimp Culture in Japan, Association for International Technical Promotion, Tokyo, Japan.

Sinha, V.R.P. et al., 1991. : Aquaculture Productivity. Oxford and IBH Publishing, New Delhi.

Srivastava, U.K. and Dharma Reddy, M. 1983 : Fisheries Development in India, Concept Publishing Company, New Delhi,

Srivastava, U.K. (Ed.) 1987 : Brackishwater Aquaculture Development in India, Concept Publishers, New Delhi.

Srivastava, U.K., et al. 1984 : Inland Fish Marketing in India-Freshwater Culture Fisheries, Concept Publishers, New Delhi,

Subramanian, S.R. 1987 : Readings in Intergrated Rural Development, Oxford and IBH, Publishing Company, New Delhi.

Subba Rao, N. 1986. : Economics of Fisheries : A Case Study of Andhra Pradesh, Daya Publishing House, New Delhi.

Subba Rao, N. 1988. : Mechanisation of Marine Fishermen : A Case Study of Vishakhapatnam, Norther Book Centre, Delhi.

Subba Rao, N. 1989. : Fisheries Development and Management in India, 1785-1986— A Bibliography, Northerna Book Centre, Delhi.

Trivedi, K.K. (Ed.) : Aquaculture Research Needs for 2000 A.D., Oxford and IBH Publishing Company, New Delhi, 1988.

Vasanth Desai 1983. : A Study of Rural Economics-A Systems Approach, Himalaya Publishing House, Bombay.

World Bank 1975. : Rural Development Sector Policy, The World Bank, Washington, U.S.A..

Younis, A.S. and Donaldson, G. 1982 : Fishery Sector Policy Paper, the World Bank, Washington, USA.

ARTICLES

Angell, C.: "Small is beautiful : On Small Scale Private Sector Shrimp and Prawn Hatcheries in India and Bangladesh", *Bay of Bengal News*, Issue No. Dec., 1993, pp. 20-22.

Anon: "Failure of Shrimp Fishery along the West Coast", *Sea Food Export Journal*, XI (IV), 1984, pp. 7-11.

Ashok Kumar, D. : "Royyalatho Videsi Currency Vindu" *Andhra Boomi*, 18th July, 1993, p. 1.

David James : "The Future for the fish in Nutrition", *Ino-Fish Marketing Digest*, Np. 4/84, No. 27-2012 MC (P) 155/84, KDN 0869/84, 1984, PP. 41-45.

Dhondyal, S.P. and Singh, G.N. : "Benefit Cost Ratios in Fish Culture", *Indian Journal of Agricultural Economics*, Vol. 23 (4), 1968, pp. 2399-242.

Ganapathy, R.P. : "Prawn Seed Requirement and its Production for Farming in India", *Sea Food Export Journal*, Vol. XV, No. 12, Dec., 1983.

Gopalan, C., et al. : "Nutritive Value of Indian foods", *National Institute of Nutrition, Indian Medical Research*, Hyderabad, 1976.

Imre Casavas : "The Impact of Aquaculture on Shrimp Industry", *Info-Fish Internatlonal, No. 1/93, Jan./Feb.*, Issue No. 127 20212, MC (P), No. 965/5/93, pp. 78/10191, 1993, pp. 47-48.

Jhingran, V.G. : "Scope and Role of Inland Aquaculture in India's Economy, *Indian Farming*, Vol.XXV, No. 6, 1975.

Kannan, R. : "Aquaculture Gains Importance", *The Hindu, Survey of Indian Industry*, 1993, Madras, P. 401.

Madhavilatha, P. and Anjaneyulu, P.S.R. : "A Study on the Input Costs and Earnings of spawn production from en ploying Chines Type Hatcheries in Krishna District", *Aquaculture Economics*, Asian Fisheries Society, Indian Branch, Mangalore, 1992, pp. 43-46.

Padmavathi, P., and Anjaneyulu P.S.R. : "A Study in Cost and Earnings of three carp culture ponds with different managerial practices in Krishna District of Andhra Pradesh", *Aquaculture Economics*, Asian Fisheries Society, Indian Branch, Mnagalore, 1992.

Pradeep, K. Katiya and Ravish Chandra: "Fish Marketing Efficiency : A Case Study of Allahabad Fish Marketing", *Fishing Chimes*, 1990, pp. 21-29.

Raju, D.S.N., et al. : "Marketing of Inland Fish from West Godavari, Andhra Pradesh, Problems and Policy Issues", *Indian Journal of Agricultural Marketing*, Vol. 3 (1), 1989, pp. 74-77.

Ramana, et al. : "Fish Marketing in Coastal Mandal-A Micro Analysis", *Indian Journal of Agricultural Marketing*, Vol. 4 (2), 1990, pp. 198-208.

Saxena, B.S. : "Net Economic Yield and the Problem of Over Fishing with particular reference to Shrimp Fishery of Kerala Coast", *Sea Food Export Journal*, Vol. 4 (1), 1972, pp. 43-50.

Silas, E.G.: "Strategies for development of aquaculture in Andhra Pradesh", *Fishing Chimes*, May, p. 14.

Subba Rao B.V. and Chowdry, K.R. : "A Study of Marketing Factors In Influencing Prices of Inland Fish in Selected Markets", *Indian Journal of Agricultural Marketing*, Vol. 2 (2), 1988, p. 170-175.

Subramanyam N.S. and Anjaneyulu, P.S.R. : "Marketing of carps from Kolleru area- Andhra Pradesh, *Aquaculture Economics*, Asian Fisheries Society, Indian Branch, Mangalore, 1991, pp. 73-76.

Subramanyan, M.V. : "Boom along the Cost", *The Hindu*, 18th July, 1994, p. 25.

Sunder, N. Sud. : "Slow Development of Inland Fisheries", *The Indian Express*, 21st Feb., 1974, pp. 7-8.

Tapash Ganguly : "In deep sea waters", *The Week*, Vol. 12, No. 3, 1994, pp. 44-46.

Upadhaya, S.N. and Dutha, A.: "Economics of Pisiculture Operations in the rural areas of Assam", *Fishing Chimes*, May, 1993, pp. 47-50.

Vijai Devi Singh, and Sampath, V. : "Present Status and Future Strategy for Brackishwater Aquaculture Development in India, *Fishing Chimes*, May, 1992, pp. 3-11.

REPORTS :

Agriculture and Co-operation, Government of India, New Delhi, Director of Fisheries, Government of Karnataka and Institute of Coastal Engineering for Fisheries, Bangalore. Proceedings of National Workshop on Brackishwater Prawn Farming for higher production, 23-25, Oct., 1989, Bangalore.

Association of South East Asian Nations, National Coordinating Agency of the Philippines, Ministry of Foreign Affairs, Manila, Philippines. *Manual on Pond Culture of Penaeid Shrimp*, A Project of ASEAN with the assistance of FAO/ UNDP, South China Sea Fisheries Development and Coordinating Programme, 1978.

Central Institute of Fisheries Research Institute. *All India Coordinated Research Project on Brackishwater Fish Farming*, Draft Final Report, (CIFRI), Barrackpore, 1985.

Centre for Inland Fisheries Research *Progress and Achievements in Aquaculture*, Information Pamphlet, Barrackpore, Orissa, 1981.

Commissionarate of Fisheries, Government of Andhra Pradesh *Freshwater Fish Culture, Information Pamphlet*, Hyderabad, 1987.

Cook, H.L. Problems in Shrimp Culture in South China Region, South China Sea Fish Development and Coordinated Programme, SCS/77/WP/140, Manila, 1976.

Dixitulu, J.V.H. Culture Fishery Development in Krishna District of Andhra Pradesh- A Case Study, Agro Climatic Regional Planning Unit, Planning Commission, Indian Sardar Patel Institute of Economic and Social Research, Ahmedabad, 1990.

FAO/World Bank Cooperative Programme. *India : Brackishwater and Inland Fisheries Project*, Staff Appraisal Report, Report No. 70/91, CP-IND, 85 1991.

George, P.C. *Fisheries Planning and Development in India, Proc.-Indo-Pacific, Fisheries Council*, 16th Session, Section-3, Bangkok, 1974.

George, P.C., et al. Fishery Resources of Indian Economic Zone (in) Souvenir issued on the occasion of Silver Jubilee Celebrations of Integrated Fishing Project, Cochin. October, 1977.

Indian Council of Agricultural Research, Central Institute of Fisheries Education Course Manual on Short term Training Programme in Management of Brackishwater Fish and Prawn Farmers, Kakinada, 16-30th Sept., 1991.

Kesteven, G.L. Shrimp Culture in Asia and Far East, A Preliminary Review, Proc. Gulf-Carrib-Fish Institute, 10, 49-68, 1957.

Ministry of Agriculture and Irrigation *Report of the National Commission on Agriculture Part VIII Fisheries*, Government of India, New Delhi, 1976.

Michaiel, B. New, et al., *Technical and Economic Aspects of Shrimp Farming*, Proceedings of the Aqua-tech, 1990 Conference, June 11-14, Kaulalumpur, Malaysia, 1990.

Marine Products Export Development Authority **INDAQUA**-Aquaculture Exposition, Souvenir, Madras, Jan. 27-29, 1995.

STAFF REPORTS :

Staff Report "A. P. Gold Mine for Prawn Culture", *Deccan Chronicle*, 15th July, 1993, p. 16.

__________, "Shrimps : The Coming Boom", *Business India*, Aug. 2-15, 1993, p. 117.

__________, "Dhanarasulanelavulu, Royyala Cheruvulu", *Andhra Jyothi*, 23rd Sept., 1993, p. 1.

__________, "Vithanala Dashalone Velandhi Mandiki Upadhi., *Udayam*, Nov. 20, 1993, p. 12.

__________, "Avakasalu Aparimitham, Cheyyutha Parimitham., *Udayam*, Nov. 20, 1993.

__________, "Law to regulate Aquaculture Farms", *The Hindu*, Oct. 12, 1994, p. 3.

__________, "Royyalade Andevesinacheyyi", *Udayam*, Jan. 27, 1994.

Yap, W.G., Apud, F.D. and Primavera, J.H. Manual on Prawn Farming, Aquaculture Extension Manual No. 5, Issue No. 0115-5369, Aquaculture Department, South East Asian Fisheries Development Centre (SEAFDC), Philippines, 1979.

INDEX